BEHIND THE SCENES
AT GALILEO'S TRIAL

BEHIND THE SCENES AT GALILEO'S TRIAL

Including the First English Translation
of Melchior Inchofer's *Tractatus syllepticus*

RICHARD J. BLACKWELL

University of Notre Dame Press
Notre Dame, Indiana

Manufactured in the United States of America

Behind the Scenes at Galileo's Trial was designed by Jane Oslislo;
composed in 10/13.3 Fairfield Light by Four Star Books;
printed on 50# Nature's Natural (50% PCR) by Thomson-Shore, Inc.

Library of Congress Cataloging-in-Publication Data

Blackwell, Richard J., 1929–
    Behind the scenes at Galileo's trial : including the first English translation of Melchior Inchofer's Tractatus syllepticus / Richard J. Blackwell.
        p.   cm.
    Includes bibliographical references and indexes.
    ISBN-13: 978-0-268-02201-3 (acid-free paper)
    ISBN-10: 0-268-02201-1 (acid-free paper)
    1. Galilei, Galileo, 1564–1642—Trials, litigation, etc.   2. Religion and science—History—17th century.   3. Inchofer, Melchior, 1585?–1648. Tractatus syllepticus.
4. Catholic Church—Doctrines—History—17th century.   5. Inquisition—Italy.
I. Inchofer, Melchior, 1585?–1648. Tractatus syllepticus. English.   II. Title.
QB36.G2B573   2006
520.92—dc22

                                                              2006016783

∞The paper in this book meets the guidelines for permanence and
durability of the Committee on Production Guidelines for Book Longevity
of the Council on Library Resources.

In Memory of My Beloved Wife
Rosemary Gallagher Blackwell
1930–2002

"Love is as strong as Death"
—Song of Songs 8:6

CONTENTS

Given the enormous size of the literature on the Galileo affair, making an addition to that collection calls for some justification. In the present case the reason is that there is an elaborate theological examination of the judgment against Galileo at his trial which quite likely was requested by the pope at the time, Urban VIII, but whose title has been only very infrequently mentioned by Galileo scholars. Moreover its central arguments have remained unexamined and unknown in English (except for two articles by William R. Shea [1984] and Thomas Cerbu [2001]), even though it presents an inside picture of the theological point of view operating during the trial for at least one, if not more, of its main participants. The purpose of this book is to present this material and an analysis of it as a new component of our knowledge of the Galileo affair.

This treatise, entitled *Tractatus syllepticus,* was written by Melchior Inchofer, S.J., whose judgment of the orthodoxy of Galileo's *Dialogue* had been requested earlier by the Holy Office and was then incorporated into the proceedings of the trial. At the time, Inchofer's judgment was the most detailed and harshest argument against Galileo's book. His later *Tractatus* is published here for the first time in English in appendix 1.

The additional sources included in the other two appendices have also been translated by the present author, as have the translations elsewhere in the book unless indicated otherwise in the notes. In all the translations any material enclosed in square brackets has been added by the translator for purposes of clarification of meaning or identification of sources. This occurs most frequently in the translation of the *Tractatus* because the Latin text of the Bible used by Inchofer in writing his *Tractatus* follows the Septuagint edition in numbering the Psalms and various other books of the Bible; in cases where the Septuagint and modern editions do not agree, the modern edition is specified in

square brackets. I have expanded abbreviated names and titles without identifying such additions to the text in order to preserve the readability of the text.

Chapter 1 is an expanded version of a paper initially entitled "Galileo's Trial:
A Plea-Bargain Gone Awry?" which was delivered at a conference entitled "Galileo and the Church" at the University of Notre Dame, 18–20 April 2002. Several presentations from that conference have been collected in McMullin (2005).

My thanks are extended to my graduate students and fellow Galileo scholars, especially Ernan McMullin, Annibale Fantoli, and George V. Coyne, S.J.,
who have researched, illuminated, and discussed the Galileo affair for me over
so many years. The shortcomings are, of course, my own. And I also must express my special appreciation to Dr. Ronald Crown and the staff members of
the Vatican Microfilm Library at Saint Louis University for their invaluable assistance in obtaining sources and tracking down references, and to Dr. Matt
Dowd, my very helpful editor at the University of Notre Dame Press.

May 2004   Saint Louis University

Ave, lector. It may perhaps be helpful for you to focus on the main themes of this book if I say a few words first about how it came to be written. My interest over the years in Galileo's clash with his Catholic Church has always drifted toward the issue of trying to understand the theological rationale behind the Church's decision to condemn Copernicanism and to bring Galileo to trial. Or, as I put it in the introduction to an earlier book (Blackwell, 1991), I have attempted to see all this not only as an episode in the history of science but just as importantly as an event in the history of theology and religion.

With this goal in mind the first area up for examination in my earlier book was the complex of events and documents centering on the year 1616, already widely studied by Galileo scholars. In that year, for complicated reasons, which are only sketched in chapter 1, the Catholic Church unfortunately decided to condemn Copernicanism as "false and completely contrary to the sacred scripture." This was a general decree of condemnation applying to all members of the Church and did not specify Galileo by name. Thus the heterodoxy of Copernicanism appears to have been an already settled question by 1616, long before Galileo published his classic *Dialogue Concerning the Two Chief World Systems,* which in 1632 quickly became an object of suspicious attention as to whether it violated the Decree of 1616. The general presumption has long been established that the theological thinking and rationale embedded behind the condemnation of heliocentrism in 1616 applied without qualification at Galileo's trial.

Perhaps that is the reason why a close reading of the records of Galileo's trial reveals no direct discussion of the Bible and its many passages referring to astronomical matters, even though those were at the foundation of the charges against him. Should we thereby conclude that the church officials and their advisers at the trial had the same theological views on these issues as did their predecessors in 1616, and especially the views of Cardinal Bellarmine, who dominated the discussions leading up to the decree of condemnation?

There are some other peculiarities about the documents from the trial. One important example is the dramatic change of Galileo's personal attitude while the trial was still in progress. A normal courtroom procedure at the time was for the prosecutor to ask questions of the accused in a series of discrete sessions of the trial, each of which was recorded accurately, although not completely verbatim, by a court scribe. This was very similar to the contemporary procedure of taking depositions for a court. At the first session Galileo made a strong and rather confident defense of himself in accounting for how his book was written. This included a letter from Cardinal Bellarmine, since deceased, which did not say that Galileo had been personally served an injunction in 1616 to abandon his views on Copernicanism but only that he had been informed of the general Decree, which applied to all Catholics. Bellarmine's letter had previously been known to no one except Galileo, and at the first session of the trial he used it as a bold and powerful appeal to a top authority in a church that highly respected such authorities.

At the beginning of the second session eighteen days later, however, Galileo's confidence was gone. The record starts out immediately with a request from him to address the court. He then proceeds to admit his guilt, not because of deliberate intention but by inadvertence. With that the second session abruptly ended. This sharp reversal of Galileo's attitude is enigmatic. Why confess at that point? Something dramatic seems to have happened in the eighteen days separating the first and the second sessions of the trial. But what? There are some clues, as we shall see in chapter 1.

Another curiosity is a document entitled the "Summary Report." The standard court procedure at that time was that the staff members of the Holy Office would write up a summary digest of what had been found in the interrogation sessions and then send that document up to the cardinals of the Congregation of the Holy Office, and sometimes on to the pope if needed, for their adjudication of a given case. The important oddity in the Galileo court documents is that the Summary Report is inaccurate and misleading in some significant ways that result in its being not fully consistent with the court's own record of the questioning of Galileo by the inquisitors. Despite its many deficiencies as a judicial system, the Roman Inquisition had a much better history of documentary accuracy than is found in this instance. Again, why did Galileo's attitude change so abruptly? What had happened?

Independently of these issues there is also clear evidence that during Galileo's trial there were initiatives made by Vatican authorities, including Pope Urban VIII, to prepare justifications of a guilty verdict at the trial. Two Jesuits, Melchior Inchofer and Christopher Scheiner, were enlisted to write theological and scientific vindications, respectively, of such a verdict. The books that they then wrote are not part of the trial documents themselves. But they do give some insight into what internal church thinking was concerning what was happening in 1633 (chapters 2–4). More specifically for our interests, the question arises of whether or not theological thinking had become modified or changed in notable ways since the Decree of 1616. And did the Church already see some of the problems that it may have created for itself in the aftermath of the Galileo trial?

The questions I have asked here are the issues that motivated my research and writing of this book. Unfortunately we do not have enough surviving evidence to answer all these questions definitively. But a careful look does reveal some significant things that happened behind the scenes at Galileo's trial.

The Legal Case at Galileo's Trial

Impasse and Perfidy

GALILEO'S TRIAL BEFORE THE ROMAN INQUISITION IS ONE OF THE MOST frequently mentioned topics in the history of science. Although no doubt only few people have carefully read and studied the actual transcripts of the trial, most have an opinion to express about it, almost always in sympathy with Galileo. All this attention is not misplaced, for Galileo's encounter with the Catholic Church not only was clearly a major turning point in the history of western culture, but unfortunately it has also been the defining event for the stormy relationship between science and religion ever since.

As a result of all this popular attention, a rather standardized stereotype of the Galileo trial has become part and parcel of our culture. We imagine Galileo standing before his peevish judges, carefully and conclusively explaining to them why the earth must revolve around the sun rather than vice versa. Meanwhile the clerical judges are sitting, as they ponder over the pages of the Bible, quoting back at Galileo various passages that now seem quite irrelevant to the issue. The scene is heavy with inevitability. The indelible image is Galileo later on his knees, forced to denounce as false a set of ideas that both he and we know to be quite true. And as he then stands, he stamps his foot on the ground and says, "Still it moves."

Needless to say, the trial did not really happen like that at all. On the scale of the very large picture, of course, the conventional image does capture the notion that the Galileo affair is of dramatic and permanent importance because it is the paradigm case of the clash between the institutional authority of religion and the new authority of scientific reason, discovered by Galileo, which has come to define the modern era. But as we focus our vision more and more finely on the specifics of the trial, a picture emerges that is quite different, much more complex, and even more ominous than what the usual stereotype portrays.

One advantage of such a closer look, of course, is that it serves as a corrective for the misleading and oversimplified features of the customary view of the Galileo trial. But more importantly it also opens the door to a whole set of new and overlooked factors that not only explain what happened more accurately but also highlight certain dimensions of the relationship of religion to science that still run deep beneath the surface.

With this objective in mind, I present in this chapter a careful reconstruction of the course of events in Galileo's trial, based on the surviving transcripts of the proceedings. What will be revealed is a scenario that shows that a legal snag surfaced early on in the trial; that the prosecutor proposed, and Galileo accepted, a compromise (which we would now call a plea bargain) designed to resolve the snag with as little damage to both sides as possible; and that at the very last minute this compromise was sabotaged by an unknown person or persons, resulting in a stunning reversal for Galileo and the prosecutor. The net result is a much different and more disconcerting picture of the role of the Catholic Church in the affair than the common stereotype projects. The nagging concern this raises for the friends of religion is whether the motivations leading to this critical reversal in the course of the Galileo trial still operate in church circles today.

Before we begin, a methodological caveat is in order. To develop a reliable reconstruction of Galileo's trial, one cannot use contemporary legal standards for either the analyses or the evaluations of what happened. One should not expect to find our now familiar legal guarantees of due process, such as representation by competent counsel, modern standards of courtroom evidence and of the authentication of documents, cross-examination by the defense, and so forth. Using such standards, one could easily, but uselessly, show that Galileo's trial falls far short of contemporary requirements for justice. Rather, in order to reconstruct the trial one must sympathetically understand and temporarily accept the legal standards used by the church courts in Galileo's own day. It is not our purpose here to judge how just or unjust those standards may have been.

The Immediate Background of the Trial

In the spring of 1632 Galileo published a book on which he had been working for many years and which was destined to bring him both fame and tragedy. The book was entitled *Dialogue Concerning the Two Chief World Systems,* a masterpiece of Italian literature in its own right. It was modeled after Plato's dialogues, with which it has often been compared. The three speakers in this fictional dialogue are Salviati (who defends Copernicus's heliocentric astronomy in a very effective and clearly Galilean fashion), Sagredo (who open-mindedly reacts to this defense at each point), and Simplicio (who obstinately defends the Aristotelian-Ptolemaic geocentric worldview no matter what comes up.) The discussions take place over a period of four days at Sagredo's palace in Venice, and the participants finally adjourn to a gondola trip on the Grand Canal.

Galileo's decision to write this book in a dialogue format is easy to understand. It enabled him to present and evaluate separately all the evidence and arguments he could muster both for and against two rival worldviews. More importantly this format also allowed him to claim that the book as a whole was neutral, and was intended to be neutral, between the two views. Whether that neutrality was actually perceived by the reader of the book is, of course, another question. That issue played a large role in the trial. The general opinion, then as well as now, is that Salviati clearly won the debate. No one has ever claimed victory for Simplicio. And a genuine neutrality reading of the *Dialogue* would be very difficult to justify.

By the summer of 1632 Galileo's new book had caused a scandal, especially in Rome where charges of heterodoxy were immediately and widely heard. The reason for this is that the *Dialogue* appeared to be a rather direct violation of a Church regulation that prohibited all books that advocated Copernicanism. This regulation had appeared in a Decree issued by the Vatican's Congregation of the Index on 5 March 1616, recent enough for all interested parties to remember with ease when they picked up Galileo's book. The events and discussions leading up to the publication of the Decree, and their evaluation, are far too complex to attempt an explanation here.[1] Suffice it to say that the issue at hand was whether Copernican astronomy contradicted certain relevant passages in Scripture that speak in terms of an earth-centered universe. If so, it was thought that Copernicanism must be false.

Since this Decree of 1616 played a central legal role at Galileo's trial, we quote here its relevant section in full.

> It has come to the attention of this Sacred Congregation that the Pythagorean doctrine of the mobility of the earth and the immobility of the sun, which is false and completely contrary to the divine Scriptures, and which is taught by Nicholas Copernicus in his *De revolutionibus orbium coelestium* and by Diego de Zuñiga in his *Commentary on Job,* is now being divulged and accepted by many. This can be seen from the letter published by a Carmelite priest, entitled *Letter of Fr. Paolo Antonio Foscarini on the Opinion of the Pythagoreans and of Copernicus on the Mobility of the Earth and the Stability of the Sun and on the New Pythagorean System of the World,* Naples: Lazzaro Scoriggio, 1615. In this letter the said Father tries to show that the above-mentioned doctrine of the immobility of the sun in the center of the world and of the mobility of the earth is both in agreement with the truth and is not contrary to Sacred Scripture. Therefore, lest this opinion spread further and endanger Catholic truth, it is ordered that the said Nicholas Copernicus's *De revolutionibus orbium* and Diego de Zuñiga's *Commentary on Job* are suspended until corrected; also that the book of the Carmelite Father Paolo Antonio Foscarini is completely prohibited and condemned; and also that all other books teaching the same thing are prohibited, as the present Decree prohibits, condemns, and suspends them all respectively. (Blackwell 1991, 122)

The Decree's message is quite unequivocal. Heliocentric astronomy is false because it contradicts the Bible, and all future books advocating Copernicanism are prohibited and condemned in advance. Of the three offenders mentioned by name, only Foscarini was still living in 1616. In fact, his recently published *Letter,*[2] mentioned in the Decree, provided an ideal occasion for the Holy Office to come to the decision announced here. It is likely that Galileo was the real target of the Decree. But perhaps because of his recently acquired international reputation as an astronomer and because of the prestige of his patron, the Grand Duke of Tuscany, Galileo is not mentioned by name in the Decree, nor is his *Letters on Sunspots* (1613) in which he explicitly advocated the Copernican view in at least two places. Nevertheless he could hardly have missed the impact of the last sentence, which understandably came to be read in the summer of 1632 as a condemnation of the *Dialogue*. There were indeed strong grounds to charge Galileo with a violation of the Decree of 1616.

In August of 1632 the Holy Office ordered that publication of the *Dialogue* be suspended, sales halted, and unsold copies confiscated. In September a Special Commission,[3] which functioned in a manner somewhat similar to a modern grand jury, was appointed to investigate the matter further. It found that Galileo had indeed defended heliocentrism in the *Dialogue,* and thus in effect had violated the Decree of 1616. Much more importantly, however, the Special Commission also uncovered in the files of the Holy Office a memorandum, previously known only to a very few, which stated that on 26 February 1616, at a meeting at Cardinal Bellarmine's residence in Rome, Galileo had been served an injunction by the Commissary General of the Holy Office in regard to the issue of Copernicanism. The specific injunction was that Galileo was ordered to abandon Copernicanism, "nor henceforth to hold, teach, or defend it in any way, either verbally or in writing." Given this very broad wording, Galileo's *Dialogue* appeared to be such a clear violation of the injunction that a trial became inevitable. It should also be mentioned that the Holy Office's memo containing the injunction came as quite a surprise even to Pope Urban VIII, who thereafter remained furious with Galileo for supposedly concealing it from him during earlier and friendlier discussions about Galileo's work. Since the 1616 injunction later became the centerpiece legal document at Galileo's trial, we quote it here in full.

> At the Palace, the usual residence of the aforenamed Cardinal Bellarmine, the said Galileo, having been summoned and standing before His Lordship, was, in the presence of the Very Reverend Father Michael Angelo Seghizzi de Lauda, of the Order of Preachers, Commissary General of the Holy Office, admonished by the Cardinal of the error of the aforesaid opinion and that he should abandon it; and later on [*successive ac incontinenti*] in the presence of myself, other witnesses, and the Lord Cardinal, who was still present, the said Commissary did enjoin on the said Galileo, there present, and did order him (in his own name), the name of His Holiness the Pope, and that of the Congregation of the Holy Office, to relinquish altogether the said opinion, namely, that the sun is in the center of the universe and immobile, and that the earth moves; nor henceforth to hold, teach, or defend it in any way, either verbally or in writing. Otherwise proceedings would be taken against him by the Holy Office. The said Galileo acquiesced in this ruling and promised to obey it. (Langford 1966, 92)

Modern scholars have shown that this Holy Office memo of 26 February 1616 contains numerous irregularities, but they are far from agreement as to how the memo should be interpreted. First it is some sort of a summary by a clerk of what happened at the meeting, but is not the official document that should have been in the Holy Office files and that would have legally stated the results of the meeting at Bellarmine's residence. Further the memo is not properly signed or witnessed. Some have argued that it was forged in 1632 to trap Galileo, but that view has been abandoned after paper, ink, and handwriting tests show that it was written in 1616. Some say that it is a 1616 forged substitute for the missing proper document, but that seems highly unlikely. Unfortunately how the memo was generated simply cannot be determined by the presently available evidence. The memo is also crucially ambiguous at its main point ("and later on" [*successive ac incontinenti*]), since we cannot tell whether Galileo was or was not given time to react to Cardinal Bellarmine's admonition before he was served with the injunction by the Commissary General. If he was not given an opportunity to reply to Bellarmine, then the injunction would have been illegal, since the pope's specific instructions for the meeting ordered the injunction as a second step only in the case that Galileo refused to abandon Copernicanism.

It is incredible that the central document in Galileo's trial, which was to have such enormous consequences, is so full of legal, textual, and conceptual problems. Further we simply do not know what the prosecutors in the Galileo trial may have known or thought specifically about the difficulties mentioned above. However they clearly were concerned in at least a general way about the status of the Holy Office memo, as we shall soon see in detail.

At any rate in light of the Decree and the Holy Office memo of 1616, Galileo was ordered in October of 1632 to come to Rome for a trial. Various delaying tactics, travel difficulties due to the plague, and Galileo's perennial poor health postponed his actual arrival in Rome to February of the next year. The stage was then set.

THE THREE SESSIONS OF THE TRIAL

Galileo's trial was conducted under the auspices of the Congregation of the Holy Office, popularly known as the Roman Inquisition, which at that time was composed of ten cardinals appointed by the pope. This Congregation was charged with the responsibility of asserting Catholic dogma and safeguarding it from any attack. The day to day work was carried out by a staff of clerics, traditionally Dominicans, headed up by the Commissary General or chief prosecutor. The usual trial procedure was that this staff would carry out the interrogation of defendants and witnesses, during which a court clerk would write down a consecutive, but not necessarily verbatim, account of the questions and answers. Immediately after each session the accused was asked to read and sign the clerk's account. Galileo's signature appears at the end of each deposition in his trial, thus increasing our confidence in the reliability of the documents.[4] After all the sessions were completed, the Inquisition staff composed a summary report of the proceedings and sent it up to the Congregation of cardinals, who would then either make the decisions, subject to the pope's approval, or else pass the matter on to the pope for his decision.

To understand the context of the documents it is important to realize that, in its dealings with individuals, the functions of the Inquisition were thought of as being primarily religious in character. The guiding purpose was to save souls by offering forgiveness of sins to the accused. For example, in Galileo's sentence he was "absolved from his deficiencies" if he accepted the judgment. The juridical proceedings were conceived of as instrumental to that end, the whole process having the atmosphere of a religious penance service.

Another characteristic feature was that everything was shrouded in the strictest of secrecy. The members of the Holy Office would never publicly say or write anything about a past or present case under pain of severe punishment. As a result there are, for example, no known comments about the 1616 deliberations against Copernicanism or about the earlier Bruno trial anywhere in the large volumes of personal papers left by Cardinal Bellarmine, and one would not expect to find any. The only records kept were the files of the Holy Office itself, which, whatever its deficiencies may have been, encouraged detailed, accurate, and legally proper documentation of its activities. Their basically reliable documents on the Galileo trial have been preserved, and were finally made public in 1880.

The court procedures were governed by published ecclesiastical legal manuals, which were standard for the times.[5] Part of the procedure was to imprison the accused for the entire duration of the trial, thus assuring, through additional procedures such as intimidation, silence about trial developments from that quarter. As a result the Inquisition came in time to manage its own prison system, which was also used, of course, to house convicted defendants.

All executions were handled by the civil authorities, lest there be any blood on the church. This is the origin of the ominous shadow behind the phrase "handed over to the civil arm."

Galileo's trial began on 12 April and ended on 22 June 1633. In a virtually unprecedented move he was not imprisoned during most of the trial. Rather he was allowed to live, under a promise of silence, at the Villa Medici, the Tuscan ambassador's residence in Rome adjacent to the Borghese Gardens, except for a critically important period of eighteen days (12–30 April) when he was held in comfortable quarters at the Dominican Convent of Santa Maria Sopra Minerva in the Piazza Minerva, the usual site of the hearings conducted by the Holy Office. Fr. Carlo Sinceri, the Proctor Fiscal, conducted the actual interrogations, under the supervision of Fr. Vincenzo Maculano da Firenzuola, O.P., who was the Commissary General. The interrogations were completed after three sessions (12 April, 30 April, and 10 May). A few weeks later the summary report was sent to the Congregation for judgment. Pope Urban VIII's decision is dated 16 June, and Galileo was sentenced six days later.

The First Session

When one looks closely at the depositions from the trial, an extraordinary story is revealed. The first session began with the usual preliminaries of identification. Galileo was then asked the standard question of whether he knew or could guess why he had been summoned by the Holy Office. The expected pro forma reply to this would have been a simple "no," lest one might give reason to suspect a guilty conscience. But Galileo more boldly replied that he imagined that it was because of his book, the *Dialogue,* for he and his publisher had received an order from the Holy Office to cease publication and to send the original manuscript to Rome. The next few questions established the fact that Galileo was the author of the book and of everything in it, and determined when and where it was written.

The first legal hurdle. At this point the interrogation took an ominous turn, which Galileo must have anticipated and recognized as such, judging from the astuteness of his replies. The question was whether he was in Rome in 1616, and why. He answered that he came to Rome that year of his own volition to learn what he was allowed to maintain about Copernicanism. He said he was informed by Cardinal Bellarmine of the soon to be published Decree of 1616 to the effect that heliocentrism could not be held absolutely, but only suppositionally,[6] since it was contrary to the Scriptures. This was the same point of view explained in Bellarmine's Letter to Foscarini of 12 April 1615,[7] in which Galileo is mentioned by Bellarmine by name as sharing precisely that view, and which Galileo quoted verbatim on this point. Appealing to Bellarmine himself for his defense was an effective move.

But Maculano was not satisfied. One has the image that he held in his hands the previously mentioned disputed memo of the Holy Office, which said much more. Perhaps Galileo sensed that he had something more damaging to use, although of course the strict secrecy rules were such that Galileo never personally read that memo, either before, during, or after the trial. The questioning became more insistent. What was decided and told to Galileo at his meeting with Cardinal Bellarmine at his residence on 26 February 1616? We are now at the dramatic highlight of the trial. Galileo repeated that the order from Bellarmine was that Copernicanism was contrary to the Scriptures and thus could not be held absolutely, but only suppositionally. He then produced a copy of the following letter given to him by Cardinal Bellarmine, dated exactly three months after the meeting, adding that the original of the letter was in safe-keeping in Rome. The letter reads:

> We, Robert Cardinal Bellarmine, hearing that it has been calumniously rumored that Galileo Galilei has abjured in our hands and also has been given a salutary penance, and being requested to state the truth with regard to this, declare that this man Galileo has not abjured, either in our hands or in the hands of any other person here in Rome, or anywhere else as far as we know, any opinion or doctrine which he has held; nor has any salutary or any other kind of penance been given to him. Only the declaration made by the Holy Father and published by the Sacred Congregation of the Index has been revealed to him, which states that the doctrine of Copernicus, that the earth moves around the sun and that the sun is stationary in the center of the universe and does not move east to west, is contrary to Holy Scripture and therefore cannot be defended or held. In witness whereof we have written and signed this letter with our hand on this twenty-sixth day of May, 1616. (Blackwell 1991, 127)

Maculano must have been shocked. He, of course, had had no way of knowing about this letter beforehand. And Galileo had produced it in court even before the prosecution brought up the Holy Office's memo, the cornerstone of the case against Galileo. The shock obviously came from the fact, which would have been evident to the prosecutor when he compared the two documents, that they were flatly inconsistent. Bellarmine's order to Galileo was simply "not to hold or defend Copernicanism," period. But the Holy Office memo says that the Commissary General, not Bellarmine, issued an injunction to Galileo, and it said that he could not "hold, teach, or defend it in any way, either verbally or in writing." This wording would have denied permission to Galileo to deal with Copernicanism even "suppositionally," the essence of Bellarmine's advice to him. As accounts of the same meeting, this will not do. The prosecutor's key evidence had been trumped even before it was introduced.

It is tempting to think, and perhaps even probable, that Galileo somehow had already learned beforehand of the damaging and previously secret memo from the Holy Office file, and that he came prepared to dramatically introduce his counterletter from Bellarmine to defend himself against it. For in September of 1632 the Special Commission had discovered and subsequently informed Urban VIII of the injunction memo, of which even the pope had been previously unaware. After that the content of the memo was no longer restricted to the files of the Holy Office. At any rate it is certainly true that, recalling the events of 1616, Galileo brought the Bellarmine letter correctly anticipating what the focus of the interrogations would be. All these years he had kept that letter in his vest pocket, as it were, as insurance against a moment like this. One could not have a better example of the wisdom of requesting official summary documents after important oral agreements have been reached.

Recovering somehow from this most unexpected development, Maculano still pressed the matter. Were there any other witnesses present, and had anyone besides Bellarmine issued an injunction? Galileo replied,

> there were some Dominican Fathers present, but I did not know them nor have I seen them since. . . . As I remember it, the affair took place in the following manner. One morning Lord Cardinal Bellarmine sent for me, and he told me a certain detail that I should like to speak to the ear of His Holiness before telling others; but then at the end he told me that Copernicus's opinion could not be held or defended, being contrary to the Holy Scripture. I do not recall whether those Dominican Fathers were there at first or came afterward; nor do I recall whether they were present when the Lord Cardinal told me that the said opinion could not be held. Finally, it may be that I was given an injunction not to hold or defend the said opinion, but I do not recall it since this is something of many years ago. (Finocchiaro 1989, 259)

These remarks are quite mysterious. What was the information from Bellarmine intended for the pope's ear only? Was it something very personal (Galileo had been a close friend of Urban VIII, a fellow Tuscan, for over twenty years), something very compromising for some high official who might otherwise use it to Galileo's disadvantage, or something quite embarrassing to the church? And why did he say this? How could it help his defense? Did the pope ever get Galileo's private message? There is nothing in the surviving documents to suggest answers to these questions. This odd request is not included in the summary report of the interrogations, even though its author would have had this information in front of him at the time. It is likely that nothing further happened on this score.

Another mysterious point is Galileo's last comment, slightly amplified later, that it was not impossible that he was given an injunction by someone else, but that he does not remember it after all these years. Why make such a concession, however slight it might be? Was Galileo honest here, or was he conveniently forgetting some damaging specifics with the plan of relying on Bellarmine's letter for his account of what really happened?

At this point Maculano seized on Galileo's last comment to introduce finally the Holy Office memo about the disputed meeting. He informed Galileo that there indeed was such an injunction, given before witnesses, and he read to Galileo the stronger wording of the memo: "neither to hold, teach, or defend it in any way, either verbally or in writing." Does Galileo remember that wording? His reply again was that this may have been said, but that he does not recall it after so many years, during which he has relied on Bellarmine's letter for what the order was. At this point the critical interrogation about the events of 1616 ended.

The second legal hurdle.   The remaining three questions put to Galileo at the end of the first session dealt with the issue of the "imprimatur" ("let it be published") of the *Dialogue,* which was an ecclesiastical license to publish. This license was not an approval of the content of a book as such, but only certified that it does not contain anything contrary to Catholic faith and morals. After the advent of printing and the concerns for orthodoxy raised by the Reformation, this ancient church practice was codified into ten rules by the Council of Trent, and the first *Index librorum prohibitorum* appeared in 1564. Since the 1616 condemnation of Copernicanism was issued by the recently constituted Congregation of the Index, Galileo's book unquestionably would need such an imprimatur.

Regarding this issue Galileo was asked first whether he had sought permission to *write* his book. His answer was "no," because his purpose was not to hold, defend, or teach Copernicanism, but to refute it, a startling claim for any one who has actually read the *Dialogue*. But of course, if he had said the opposite, he would have condemned himself at the trial by his own words. At least he showed that he was keenly aware of the charge in the injunction memo.

Secondly he was asked if he had sought permission to *publish* the book. His long answer was not only "yes," but also that he had been granted not just one imprimatur but two. Galileo had gone to Rome in 1630 to request an imprimatur from Niccolò Riccardi, O.P., the Master of the Sacred Palace, who normally handled such matters. Galileo understood that Riccardi had full powers to approve, reject, or modify the book. After some delays he granted approval for the book to be published in Rome, but with the qualification that he review the final copy, especially its Preface and conclusion. After Galileo returned to Florence, long delays continued in Rome, indicating that Riccardi was under pressure and uncomfortable about the matter. Meanwhile that fall the plague broke out in central Italy, and it was too dangerous not only for Galileo to return to Rome but also for him to send the manuscript, which might be lost, damaged, or even burned because of the quarantine. So Riccardi ultimately agreed to let the inquisitor in Florence make the final decision as long as his initial requirement was met, that is, that the preface and conclusion (which contained the famous papal objection based on divine omnipotence) be sent to him first. As a result the *Dialogue* was finally published in Florence in 1632 with an imprimatur from Riccardi in Rome and another from Clemente Egidi, O.F.M., in Florence.

Thus it happened that the book was published with a double imprimatur. The legal quandary that this presented to the prosecutor at the trial was obvious. How could the Church protect its good name and image if it were to condemn Galileo for publishing a book that the Church itself had recently approved?

Perhaps Galileo had obtained the imprimatur fraudulently. So the last question at the first session was whether Galileo had informed Riccardi of the injunction imposed upon him in 1616. Galileo's answer was "no," for the curious reason again that his purpose was to refute Copernicanism, not to hold or defend it. One might have expected him to say that he was guided by Bellarmine's letter about that meeting and had as a result forgotten anything else by 1630.

With this the first session ended. Galileo then signed the deposition, was sworn to silence, and was ordered detained in the quarters of the Holy Office, for what turned out to be the next eighteen days.

The Plea Bargain

Maculano probably did not sleep well for the next few nights. His case had fallen apart because of the course of events in the first session. In the large picture we see that both Galileo and the prosecutor had entered the trial with a document in hand, unknown to the other side, which each thought would conclusively settle the matter in his favor. But the two documents, when considered together, were so irreconcilable as to result in an impasse. If Maculano was upset, genuine fear of the Inquisition must have descended on Galileo, who by now must also have realized what the situation was.

Unfortunately there was no clear way to remove the impasse. One word from Bellarmine would have been enough for the court to determine which document presented the true account of the critical meeting in 1616. But he had died in 1623. Michelangelo Seghizzi, O.P., the Commissary General in 1616 who supposedly issued the injunction, was also dead. There were no witnesses to call upon for a resolution.

Meanwhile pressure continued to build on the prosecutor. The three member Special Commission of the previous September had been reconvened with the specific charge to examine the *Dialogue* to determine if it "holds, teaches, or defends in any way" the following two claims: (1) that the sun is at rest in the center of the universe and (2) that the earth moves around the sun. It is interesting to note that these same two claims originally appeared in the deliberations of the Holy Office in 1616 and were then judged to be false and contrary to the Scriptures. In 1633 the Holy Office was closely following its own internal documents.

On 17 April the three members of the Special Commission submitted separate reports,[8] agreeing unanimously that the *Dialogue* had indeed violated the injunction as stated in the Holy Office's memo of 1616. The report submitted by Fr. Melchior Inchofer, the Jesuit member of the Commission, was particularly vigorous and meticulous in its indictment of the *Dialogue*. On 21 April the Congregation approved the judgments against Galileo's book by the members of the Special Commission. On the next day Maculano wrote to Cardinal Barberini, in a letter just recently discovered, to suggest a speedy settlement of Galileo's trial, both because of the Congregation's decision on the book and because of the deterioration of Galileo's health while he was under the confinement of the Holy Office. The relevant portion of the letter reads:

Last night Galileo was afflicted with pains which assaulted him, and he cried out again this morning. I have visited him twice, and he has received more medicine. This makes me think that his case should be expedited very quickly, and I truly think that this should happen in light of the grave condition of this man. Already yesterday the Congregation decided on his book and it was determined that in it he defends and teaches the opinion which is rejected and condemned by the Church, and that the author also makes himself suspected of holding it. That being so, the case could immediately be brought to a prompt settlement, which I expect is your feeling in obedience to the Pope.[9]

The first thing to note in this letter is that in the eyes of the prosecutor the case against Galileo had in effect already been decided the previous day when the Congregation voted to accept the three reports from the Special Commission without any qualifications. Galileo was guilty of "teaching, defending, and holding" Copernicanism. That matter was settled. The issue now became how to implement this decision.

Secondly, faced with a complex situation after the first session of the trial, Maculano[10] and Francesco Cardinal Barberini, a nephew of the pope and a member of the Congregation, jointly decided to try to end the case by means of a plea bargain with Galileo. For this maneuver the approval of the cardinals constituting the Congregation of the Holy Office would first be needed. So Maculano met with them to review the case to date, and to consider "various difficulties in regard to the manner of continuing the case and leading it to a conclusion." In his letter of 28 April 1633 to Cardinal Barberini, who was at Castel Gandolfo at the time, Maculano went on to describe what happened.

Finally I proposed a plan, namely that the Holy Congregation grant me the authority to deal extrajudicially with Galileo, in order to make him understand his error and, once having recognized it, to bring him to confess it. The proposal seemed at first too bold, and there did not seem to be much hope of accomplishing this goal as long as one followed the road of trying to convince him with reasons; however, after I mentioned the basis on which I proposed this, they gave me the authority.[11]

The prosecutor was now authorized to try to strike a plea bargain with Galileo. In effect, Galileo would plead guilty to some as yet unspecified minor offense in writing the *Dialogue* in return for a lighter sentence. This authority was granted after the prosecutor mentioned the "basis" of his proposal. This "basis" is not identified, but presumably it was the need to resolve the legal impasse created by the two opposing documents or the double imprimatur at the center of the interrogations of the first session. Maculano then continued in his letter to tell Cardinal Barberini that his attempt to deal with Galileo on this matter was also successful.

> In order not to lose time, yesterday afternoon I had a discussion with Galileo, and, after exchanging innumerable arguments and answers, by the grace of the Lord I accomplished my purpose: I made him grasp his error, so that he clearly recognized that he had erred and gone too far in his book; he expressed everything with heartfelt words, as if he were relieved by the knowledge of his error; and he was ready for a judicial confession. However, he asked me for a little time to think about the way to render his confession honest, for in regard to the substance he will hopefully proceed as mentioned above. (Finocchiaro 1989, 276)

Galileo had decided to admit that he "had erred and gone too far in his book." Further he was relieved and thankful for this development, which clearly implies that he also was highly concerned about how the impasse in the trial would ultimately affect him. Since the next step would be a confession in court, which would be dramatically in contrast to his assertive stance in the first session, Galileo asked for some time to think about how to make the confession credible. The decision was already made on other grounds; the publicly stated reasons for it were to be supplied after the fact. Everything would then be in place. Maculano and Cardinal Barberini are to be applauded for having enough of a sense of justice to try to arrange for the Galileo case to change into a more minor matter with more minor consequences. The prosecutor ended his letter with the hope that

> in this manner the case is brought to such a point that it may be settled without difficulty. The Tribunal will maintain its reputation; the culprit can be treated with benignity; and, whatever the final outcome, he will know the favor done to him, with all the consequent satisfaction one wants in this. (Finocchiaro 1989, 277)

So by the end of April the situation was as follows. Maculano had initiated a suggestion, which Galileo accepted, that the impasse reached at the end of the first session of the trial be resolved by what we now call a plea-bargain agreement. Galileo would "confess" to lesser offenses in his book (i.e., less than formal heresy), and the prosecution would agree to accept that as a basis to resolve the case, including a proportionally lesser penalty for Galileo. This carefully worded confession would be presented at the second session of the trial on 30 April. For reasons to be explained later, we do not know what may have been the specific lesser final judgment and penalty promised by Maculano as his half of the plea bargain, but certainly it would have been a finding of something less than formal heresy. The Holy Office had many categories of theological error short of heresy which could have been used, for example, *temeritas* or "rashness."

With this plea bargain in mind Maculano hoped to overcome the two legal hurdles that arose in the first session of the trial: (1) the two conflicting documents describing Galileo's meeting with Bellarmine in 1616 regarding what was acceptable and (2) the double imprimatur granted to Galileo's *Dialogue* shortly prior to its publication. It was a good plan, which should have worked.

The Second Session

From this point on, the character of the court sessions fundamentally changed. Interrogations to obtain information ended. The new goal was to implement the plea bargain. As a first step Galileo played his part by petitioning the court to make an opening statement at the second session of the trial. He began by saying that since the end of the first session he has been wondering whether he may actually have violated the injunction stated in the Holy Office's memo. So he reread his *Dialogue*, put aside for three years already, to see in retrospect if, contrary to his best intentions and through oversight, "some words might have fallen from my pen" that indicated disobedience on his part. He came to the following conclusion:

> Now I freely confess that it appeared to me in several places to be written in such a way that a reader, not aware of my intention, would have had reason to form the opinion that the arguments for the false side, which I intended to confute, were so stated as to be capable of convincing because of their strength, rather than being easy to answer. (Finocchiaro 1989, 278)

How could Galileo say with any honesty that his real "intention" in the *Dialogue* was to refute Copernicanism (which simply was not true)? Or on the other hand, was this rather what both sides agreed in private discussions that Galileo would say as part of the settlement, and they would simply leave it at that? Whatever be the case, he went on to say that the dialogue style of the book required that strong arguments be given for the Copernican position before he could proceed to refute them. Further, at times he became carried away with a desire to display his own cleverness in making these false views appear probable. But this was personal exuberance, not disrespect for the teachings of the Church. "My error then was, and I confess it, one of vain ambition, pure ignorance, and inadvertence" (Finocchiaro 1989, 278). While leaving, he turned back with the afterthought that he could add one or two more days of discussion to the *Dialogue* to refute the false point of view more effectively. Needless to say, the Holy Office, not wanting to encourage more trouble, paid no attention to that suggestion. The court asked him no questions about his sudden "confession," and the second session ended at that point. After signing the deposition again, Galileo was released to return to the Villa Medici, the prosecutor rather obviously being satisfied that the whole matter was now safely in hand.

The Third Session

The next prescribed step was for the court to offer the defendant an opportunity to present a defense, if he so wished. So ten days later Galileo appeared again and presented two documents: (1) the original copy of Bellarmine's 1616 letter to him and (2) a brief written defense of his actions. In the latter he said that Bellarmine's letter, which he used in the subsequent years as his guide in these matters, did not contain the stronger wording of the Holy Office's injunction, which he then understandably did not attend to. As a result he did not willingly disobey the orders given to him in 1616.

> Thus those flaws that can be seen scattered in my book were not introduced through the cunning of an insincere intention, but rather through the vain ambition and satisfaction of appearing clever above and beyond the average among popular writers; this was an inadvertent result of my writing, as I confessed in another deposition. I am ready to make amends and compensate for this flaw by every possible means, whenever I may be either ordered or allowed by Their Most Eminent Lordships. (Finocchiaro 1989, 280–81)

He concluded by asking the court to mercifully consider his declining age of seventy years, his suffering from his perennial poor state of health, and the slanders against his reputation as adequate punishment for his crimes. This appeal for leniency was most probably only a pro forma set of remarks, and not any indication of an actual penalty agreement included in the plea bargain. Galileo's role in the proposed settlement was now completed. He needed only to wait for Maculano to carry out his part of the agreement.

THE FALSE SUMMARY REPORT

Following regular procedures, the staff of the Holy Office next prepared a summary report on the trial to date, which was then passed on to the Congregation, and ultimately to the pope, for judgment. It is important to emphasize that this was an internal document of the Holy Office, and hence Galileo never saw it and had no opportunity to challenge its accuracy. The document has no date or signature.[12] The latter is easy to explain in light of the fundamentally deceptive character of the report, which required concealment of the author(s). Its official status as the legal summary of the case can be seen in the fact that, still today in the Vatican archives, this document is the first one encountered in the collection of the Galileo trial documents, which otherwise are in chronological order.

When one compares the summary report with the three depositions and the other relevant documents, one sees first that most of it is a synthesis of the three sessions of the trial. However it starts with some other background material from 1615–16, which would have previously been in the Galileo file of the Holy Office. Much more importantly one next sees that some parts of the report are accurate, other parts are deceptively misleading, some major factors in the trial documents are simply omitted, and still other parts are deliberate falsifications. No honest lawyer would have written this summary report. What we have rather is a willingness to compose a misleading and partially false document.

The summary report begins by reviewing at length a complaint received by the Holy Office in February 1615 from Fr. Niccolò Lorini, O.P., who questioned Galileo's orthodoxy then, and by implication, again in 1633. Since we have not discussed this episode earlier, a brief synopsis of it is in order.[13] On 14 December 1613 Galileo had written a letter to his scientific associate and friend Fr. Benedetto Castelli, O.S.B, in which he sketched his views about the relations between science and the Bible. The point of the letter was to put to rest an initially small dispute on this topic that had arisen in a discussion within the Tuscan royal family, who were Galileo's patrons. In February 1615 Lorini sent a copy of that letter to the Roman Inquisition with a complaint that Galileo's views in it were suspect. To make matters worse, Lorini falsified Galileo's Letter to Castelli in several places.[14] For example, regarding the Scriptures he changed "give an impression which is different from the truth" to "which, in respect to the bare meanings of the words, are false." Again he changed "Scripture does not refrain from faintly sketching its most important dogmas" to "does not refrain from perverting its most important dogmas." These falsifications are even explicitly quoted against Galileo in the summary report.

For some reason Galileo suspected such foul play, so he sent a true copy of his Letter to Castelli to his friend Msgr. Dini in Rome, along with a request that he give it to Cardinal Bellarmine, which he did, along with many other copies to others. Galileo's original letter was thereby made public knowledge. Also since at the time Bellarmine was a member of the Holy Office, he may well have added the true copy to the file on Galileo, where Lorini's complaint was lodged. This would help to explain why that complaint was soon dropped.

Yet here we have the author of the summary report of 1633 not only going back to that discredited episode, but also using Lorini's falsified version of Galileo's letter, even as he added, "Despite diligent efforts one could not obtain the original of this letter" (Finocchiaro 1989, 282). However Galileo's version of the letter was not only public knowledge since 1615, but it may well have been right there in Galileo's file. This was not even a well concealed deception; yet no one who read the report seemed to notice it, as far as we know. The summary report contained the list of Lorini's complaints of heterodoxy, plus another list of such views (i.e., "that God is an accident; that He really laughs, cries, etc.; and that the miracles attributed to the Saints are not true miracles") submitted in 1615 by Lorini's friend Fr. Tommaso Caccini, O.P. Although all this had been dismissed long ago, it made Galileo look like a veteran troublemaker who could not be trusted.

The summary report continued with more guilt by innuendo. When the Holy Office debated the orthodoxy of Copernicanism in 1616, it asked its theological experts for their opinion on these two claims: (1) that the sun is at rest in the center of the universe and (2) that the earth rotates around the sun. The theologians advised, and the cardinals of the Holy Office agreed, that both of these claims were "philosophically absurd" and that the first one was also "formally heretical," that is, it directly contradicted the words of Scripture. (It is interesting to note that the latter phrase was not used in the Decree of 1616 published shortly thereafter.) The author of the summary report then added gratuitously that these two propositions were derived from Galileo's *Letters on Sunspots*. The effective impression was that Galileo had previously defended views judged to be heretical by the Holy Office.

The remainder of the summary report deals with the content of the three trial depositions and is thus the heart of the matter. Neither the disputed Holy Office memo nor Galileo's 1616 letter from Bellarmine are quoted in full; only a few phrases from each are quoted. Nor is there any indication that these documents were added to the report as appendices for the reader to consider. What we get rather is a misleading and partially false synopsis. On the key issue of the two documents we are told that Bellarmine explained to Galileo the impact of the Decree of 1616, namely, that Copernicanism cannot be held or defended. The report also says, falsely, that Bellarmine (it clearly was not Bellarmine, but the Commissary General, Michelangelo Seghizzi, according to the Holy Office's own memo) also issued a more specific injunction to Galileo "not to hold, teach, or defend [it] in any way, verbally or in writing." In no way are we given the impression that in the documents these are two separate and inconsistent accounts of the same event. Rather the two accounts are presented as supplementary. In this scenario the weaker order is the thrust of the Decree, which applied generally to every Catholic, while the stronger order was the force of the injunction, which applied personally to Galileo. Hence he was consistently bound by both ordinances.

Given this distorted summary account, there was consequently no weight left in Galileo's defense that he later was guided only by the conditions in Bellarmine's letter, and thereby did not remember anything further about the matter. This defense is reported, but it does not have its original significance. The summary report also included a long and basically reliable account of Galileo's conduct in gaining permissions to publish in Rome and in Florence. Furthermore, unlike the excluded texts of the two key letters, the report also contained a nearly verbatim quotation of almost all of Galileo's "confession" from the second deposition. But now, taken out of the context of the plea bargain which remained unmentioned, the confession did not have the same meaning; in fact, it had become damaging to Galileo.

The summary report ended by stating what it took to be Galileo's motives:

He begged to be excused for having been silent about the injunction issued to him, since he did not remember the words "to teach in any way whatever," and so he thought the decree of the Congregation of the Index was sufficient. . . . He said all this not to be excused from the error, but so that it be attributed to vain ambition rather than to malice and deception. (Finocchiaro 1989, 285–86)

Who wrote this summary report? It could have had a solitary author. But given the complexities of pulling off such a maneuver, it is much more likely that it was the product of a group of individuals who could control the situation, and who decided for unknown reasons to ensnare Galileo. Further it is almost inconceivable that all this could have happened just at the lower staff levels of the Holy Office. More highly positioned people there, or in the Congregation itself, or even above, were almost certainly involved. But there is no known evidence anywhere to identify the author(s), or even to reasonably speculate.[15] Concealment was an essential part of this plot, massively aided as it was by the excessive secrecy of the Holy Office itself, which was being manipulated by its own rules.

Whoever it was, he knew what he was doing. By omitting all traces of the plea bargain (e.g., there was not the slightest indication in the summary report of Maculano's extrajudicial meeting with Galileo), and by very boldly distorting the documents (which were still there threatening to reveal the plot), the perpetrator(s) managed to entrap Galileo in a miscarriage of justice. With the plea bargain sabotaged, whatever specific promises Maculano had made to Galileo in late April regarding the outcome of the trial were irretrievably broken and are to be found nowhere in the surviving documents.

Meanwhile Galileo did not learn until the middle of June how badly things were going against him. In fact he was very optimistic, even confident, about the outcome, as is evident in his correspondence with numerous friends and family during May and early June. The only one at the Villa Medici who knew otherwise was Francesco Niccolini, the Tuscan ambassador, but he decided not to inform Galileo yet.

THE JUDGMENT AND THE SENTENCE

When the summary report on the trial reached the cardinals of the Holy Office, probably sometime in late May, it must have created a vigorous discussion and some disagreement. At least one of them, and probably a few others, expected a quite different report. For in fact Cardinal Barberini had had a hand in fashioning the plea bargain. So the summary report must have struck him as being an attempt to sabotage the agreement that he himself had tried to arrange. On the other hand there must also have been a faction in the Congregation who favored the summary as the basis for further action in the trial. Otherwise it is hardly credible that the report was not rejected by the Congregation as misleading, particularly after Barberini, who, after all, spoke with the authority of the pope's nephew, would have explained the situation. Put in another way, a plot to sabotage the plea bargain would almost certainly never have been even initiated, and definitely could not have succeeded, without the support of some of the cardinals of the Congregation. A likely result could have been another stalemate, this time in the Congregation itself.

But all of this is purely speculative. We have no concrete evidence at all as to what happened when the summary report was evaluated by the cardinals. But there is one small piece of indirect evidence which might indicate dissension at that level. A few weeks later, when Galileo was finally condemned, three of the ten cardinals did not sign the sentence. One of them was Cardinal Barberini.[16]

Whatever may have transpired among the cardinals of the Holy Office, Galileo's case and his sentence was finally settled personally by Pope Urban VIII. His decision on the matter, dated 16 June, was as follows:

> His Holiness decreed that the said Galileo is to be interrogated with regard to his intention, even with the threat of torture, and, if he sustains [that is, answers in a satisfactory manner], he is to abjure *de vehementi* [i.e., vehement suspicion of heresy] in a plenary session of the Congregation of the Holy Office, then is to be condemned to imprisonment as the Holy Congregation thinks best, and ordered not to treat further, in any way at all, either verbally or in writing, of the mobility of the earth and the stability of the sun; otherwise he will incur the penalties for relapse. The book entitled *Dialogo di Galileo Galilei Linceo* is to be prohibited. (Langford 1966, 150)

Added to this decree was an order that it be widely distributed and made public, especially to as many "mathematicians" as possible. No more such transgressions were to be tolerated. We do not know how Galileo first learned of the decision, but he must have been stunned. He had unexpectedly been betrayed. For the remainder of his days he expressed nothing but contempt for his judges.

Thus it was that Galileo was pronounced guilty, presumably of violating the Decree and the Holy Office's injunction of 1616, although that is not specifically mentioned in the pope's order. The next steps were routine, designed to absolve Galileo of his guilt and to punish him for it, under the model of a religious penance service.

First he was to be subjected to an interrogation, under a verbal threat of torture, to establish what his intentions were. Galileo scholars now agree that no torture occurred, nor could it have occurred, given his age and poor health, according to the rules of the Holy Office itself, and Galileo would have known this. But although the verbal threat was pro forma, the terror it was intended to evoke must have been quite real. On any account the use of torture as part of a juridical process, seen as despicable now, was not uncommon in the seventeenth century even under church auspices, as for example in the case of Thomas Campanella, O.P.

If Galileo failed the test, that is, if his intentions as revealed in the interrogations proved to be suspect, he would have been judged to be an "impenitent and obstinate" heretic, in other words, one who admitted heresy, defended it, and refused to recant. The penalties for this officially recognized category of offender were extremely severe. On the other hand, if his intentions were judged to be innocent, then a series of further steps would take place. After being sentenced, he would be required to read an abjuration, a denial of his offending views, before the cardinals of the Holy Office, since he had been judged to be guilty of "vehement suspicion of heresy."[17]

The penalty ordered by the pope was that (1) Galileo was to be imprisoned at the discretion of the Holy Office, and (2) the *Dialogue* was to be forbidden and placed on the *Index*. If Galileo were later to deal with Copernicanism again in any way, he would then be declared a "relapsed" heretic, a second time offender after a reconciliation, who was thereby considered incorrigible. The standard penalty for a *relapsus* was execution. In the light of the legal prescriptions for various categories of heretics implied in the pope's decree, those who would criticize Galileo for not resisting further after this point do not really understand the procedures and the terror of the Roman Inquisition. From here on, he had to play by their rules or face the most extreme consequences.

Why did Urban VIII make this set of decisions? Had he read the depositions of the trial? Did he know about the conflict of the key documents? What did he think about the summary report? Was he aware of, or even a party to, the sabotaging of the plea bargain? What legal reasoning led him to his judgment?

Was he motivated by other factors (for example, he was an old friend, admirer, and even encourager of Galileo ten to fifteen years earlier, but had become furious with him after he first learned in the fall of 1632 about the disputed Holy Office memo and its apparent injunction to Galileo)? How offended was the pope that his own views on the matter were spoken by the close-minded Simplicio in the *Dialogue*? Or did he fail to really give the matter adequate attention, being distracted by, and suspicious of, the political and military events going on about him over the Thirty Years War?

We will almost certainly never know the answers to these questions. The records simply are not there. Part of what happens in an organization built on a highly centralized authority, operating often in secrecy, is that the top person is too easily shielded from the light of truth (in the double sense that he may not know all the relevant factors in making his decisions and that those outside the inner circle may not know the real reasons for his actions). This is often seen as an advantage by those who think they are protecting that person and the institution, but it can also damage its credibility.

The remaining proceedings in the Galileo case were pro forma. On 21 June he was called again before the Commissary General, to be interrogated this time about his intentions. He was asked three times, in three different contexts, whether he held that Copernicanism was true. He replied that before 1616 he thought that both the Ptolemaic and the Copernican systems were open to dispute, but after the Decree of 1616, "assured by the prudence of the authorities, all my uncertainty stopped" (Finocchiaro 1989, 286). Since then he has held that the Ptolemaic view was true. His book had reviewed all the arguments on both sides with the purpose of showing that none of them were conclusive, "so one had to resort to the determination of more subtle doctrines" (Finocchiaro 1989, 287). He has not held Copernicanism to be true since he was ordered in 1616 to abandon that idea. That being said, Galileo signed his name to the deposition. Like St. Peter, Galileo denied the truth three times. Peter at least was given a free choice.

The next day, 22 June, the formal sentence was read at a full session of the Congregation of the Holy Office, after which Galileo read aloud the abjuration statement that had been prepared in advance for him. This is the scene usually called to mind by the customary stereotype of the Galileo trial. But actually the matter had been settled days earlier, no arguments were any longer to be heard, and the two documents involved merely brought the issue to a formal legal closure. It is Galileo's public humiliation at this point, of course, which carries the dramatic weight of the moment. But which side was really humiliated?

The sentence contains a long preamble stating the facts of the case, facts which we have already examined in detail. We need only mention that whoever prepared the actual text of the sentence clearly used the misleading summary report, whose language is repeated verbatim in places. The actual sentence is contained in only a few lines:

> We say, pronounce, sentence, and declare that you, the above mentioned Galileo, because of the things deduced in the trial and confessed by you as above, have rendered yourself according to this Holy Office vehemently suspected of heresy, namely, of having held and believed a doctrine which is false and contrary to the divine and Holy Scripture: that the sun is the center of the world and does not move from east to west, and the earth moves and is not the center of the world, and that one may hold and defend as probable an opinion after it has been declared and defined contrary to the Holy Scripture. Consequently you have incurred all the censures and penalties imposed and promulgated by the sacred canons and all particular and general laws against such delinquents. (Finocchiaro 1989, 291)

In the spirit of a penance service, Galileo was then told that the Holy Office was willing to give him forgiveness of his guilt if he read with a sincere heart the abjuration statement, already prepared for him, in which he would deny and curse all heretical errors, including his own. Three penalties were then listed, two of which had been stipulated earlier by the pope. The *Dialogue* was to be placed on the *Index* as prohibited, Galileo was condemned to "formal imprisonment" at the discretion of the Holy Office, and he was to recite the seven penitential psalms weekly for the next three years.

Immediately afterward Galileo, kneeling before his judges, read and then signed the abjuration statement. It was in the form of an oath to abandon Copernicanism and had the legal effect of removing the "vehement suspicion of heresy" of which he had been judged guilty. Its central paragraph reads:

> Therefore, desiring to remove from the minds of Your Eminences and every faithful Christian this vehement suspicion, rightly conceived against me, with a sincere heart and unfeigned faith I abjure, curse, and detest the

above mentioned errors and heresies, and in general each and every other error, heresy, and sect contrary to the Holy Church; and I swear that in the future I will never again say or assert, orally or in writing, anything which might cause a similar suspicion about me; on the contrary, if I should come to know any heretic or anyone suspected of heresy, I will denounce him to this Holy Office, or to the Inquisitor or Ordinary of the place where I happen to be. (Finocchiaro 1989, 292)

This is the scene, burned into the public memory, that is the core of the traditional image of the Galileo case. But the heart of the matter was the earlier event of the plea bargain gone awry, which had come back to convict him. And thus it was that this unpleasant business at the Piazza Minerva came to a conclusion.

Before leaving this final scene, we should ask ourselves whether the outcome of the trial would have been different if the plea bargain had not been blocked by the misleading summary report. The answer, of course, is, "Yes, considerably different." The two legal hurdles that arose in the first session of the trial could then have been simply left unresolved. Given Galileo's confession, there would have been no need to reconcile the differences between the two documents (i.e., the Holy Office's injunction memo and Bellarmine's letter to Galileo) describing the instructions given to Galileo in 1616. And the imprimaturs granted to Galileo's *Dialogue* in 1632 would not be as embarrassing if the book had been declared to be only objectionable but not formally heretical. And the summary report would then have needed only to collect the testimony given and to describe what the plea bargain was in specific detail.

But primarily, if the plea bargain had not been blocked, this would have had the great advantage of letting the Holy Office bring a judgment of something less that heresy (which term had *not* been used in the public Decree of 1616.) The Holy Office and the Pope had numerous categories of theological error less than heresy in its repetoire, which could have been used for a softer conclusion to the trial. If that had happened, the church would have been much better served since that would have obviated any need in later years to justify a "heresy" judgment.

At any rate the promise made by the prosecutor was in fact broken, and this may well have contributed to Galileo's strong bitterness about the trial in later years. For whatever the prosecutor had offered for his cooperation in the plea bargain, Galileo certainly would not have expected to be labeled a heretic and to see his personal freedom restricted for the rest of his life.

THE AFTERMATH

One week after the trial Galileo's sentence of imprisonment was commuted to what we would now call "house arrest" for the rest of his days, first at his small villa at Arcetri near Florence, and later in Florence after he had lost his vision in 1637. The recitation of the penitential psalms was transferred, at his request, to Sister Maria Celeste, a Carmelite nun, who was Galileo's elder daughter.

Although forbidden to deal in any way again with Copernicanism, Galileo continued his scientific work in the remaining nine years of his life. He wrote another set of dialogues entitled *Discourses on Two New Sciences* (Leyden, 1638), which pulled together and perfected his ideas going back more than three decades on abstract topics in theoretical statics and dynamics. These writings were his most substantive contributions to physics in the long run, and in effect laid the groundwork for Newton's later theoretical justification of the Copernicanism which Galileo had been forbidden to discuss.

When Galileo died in 1642, the Duke of Tuscany requested permission to construct a tomb and monument in his honor in the Church of Santa Croce in Florence. But the request was denied by Urban VIII because Galileo had caused "the greatest scandal in Christendom." He was buried instead in the basement of the bell tower. Nearly one hundred years passed before his body was moved and interred with the intended honors in the church proper, where his remains lie today adjacent to two other famous local citizens, Michaelangelo and Machiavelli. Two hundred years had to pass before the condemnation of the *Dialogue* was removed.[18] Two hundred and fifty years had to pass before the seal of secrecy was removed from the trial documents held in the Vatican Secret Archives.

If we step back now from the specifics of Galileo's trial and look at it as a whole, it is clear that some member(s) of the Holy Office, whose names remain unknown, sabotaged the plan for a plea bargain that was designed to bring the trial to a convenient close. Their misguided actions were probably motivated by a desire to serve the good of the church, although the results were overwhelmingly the opposite of that in the long run. Perhaps they still felt the shadow of the Reformation so strongly that they viewed Galileo as potentially another Luther.

Melchior Inchofer's Role
in the Galileo Affair

IN THE YEARS AFTER HIS TRIAL AND CONDEMNATION GALILEO REMAINED
convinced that his downfall had been caused by a plot against him by his enemies. Furthermore in his correspondence of those years he explicitly says in
several places that this plot was engineered by the Jesuits. His most direct comment to this effect is in his letter to Elia Diodati of 25 July 1634.

> They [my persecutors] have finally decided to reveal themselves to me. It
> so happens that at a meeting about two months ago in Rome a dear friend
> of mine had a discussion with Fr. Christopher Grienberger, a Jesuit mathe-
> matician at their College. When my affairs came up, this Jesuit spoke these
> specific words to my friend: "If Galileo had known how to keep on friendly
> terms with the Fathers of this College, he would be enjoying fame in the
> world, he would not have had any misfortunes, and he would be able to
> write freely about anything, even the motion of the earth." Thus you see
> that it is not this or that opinion which has caused the past and present
> warfare against me, but rather it is my being held in disfavor by the Jesuits.
> (Galileo 1890–1909, 16:116–17)

As a result of this and other such remarks by Galileo, his sympathetic contemporaries also occasionally spoke about a Jesuit plot against him. This same
theme has been both defended and rejected in various twentieth-century studies of the Galileo case.[1] Nevertheless it has proven very difficult to either verify
or falsify this claim.

If Galileo meant that there were a few Jesuits who strongly opposed him
as individuals, then he was clearly correct in his complaints. His sharpest Jesuit scientific opponent was Christopher Scheiner, who fell out with him in
1613, twenty years before the trial, over the issues of priority of observation and
of interpretation in regard to sunspots, and who also wrote bitter personal attacks against Galileo and his heliocentric views before and during the trial
(Scheiner 1626–30, 1651). Another Jesuit opponent was Orazio Grassi, whom
Galileo engaged in a sharp dispute for seven years (1619–26) over the proper
understanding of comets.[2] Still another was Melchior Inchofer, S.J., who, as
we shall see later, played a direct role in Galileo's trial. And of course there were
still others.

On the other hand Galileo was honored by the Jesuits in May 1611 at a
conference at their College in Rome which confirmed his telescopic observations. Furthermore some Jesuits, for example, the astronomers Giuseppe Biancani and Christopher Grienberger, are known to have largely agreed with Galileo on Copernicanism in 1616 (Blackwell 1991, 148–53). As a result one must
say that Galileo had both opponents and supporters among the Jesuits, although
after the condemnation of Copernicanism in 1616, and as the years passed on
toward the trial in 1633, and especially beyond it, the opponents were an increasingly large and loud majority.

If Galileo's complaints about a Jesuit plot mean in addition that his known
Jesuit opponents worked together to bring about his trial, as some scholars have
suggested, then that is a more difficult thesis to justify. That is a possible scenario, of course, but there is little specific evidence to justify that claim, only
very indirect intimations at best. However there is some evidence of collusion
between Scheiner and Inchofer, as we shall see later.

Lastly, if Galileo's complaints about the Jesuits are taken to mean that the
Jesuits as a corporate body had a policy to oppose him, then that is almost certainly false. The closest thing to this that one can find is the decision in 1611 by
the General of the Society, Claudio Aquaviva, S.J., that in their schools the
Jesuits must follow the teachings of St. Thomas in theology and of Aristotle
in philosophy.[3] The latter, of course, involved a commitment to a geocentric
astronomy, and thus an opposition to Copernicanism, although most Jesuit

astronomers after 1616 actually adopted Tycho's modified geocentrism rather than Aristotle's own theory. It is clear that Aquaviva's policy hampered the many Jesuit astronomers of the time who had sympathies with Copernicanism, but that is a long way from saying that there was a policy of organized opposition to Galileo. There is simply no evidence of the latter.

The result of all this is that Jesuit involvement in the Galileo affair remains as unclear today as it was then.[4] We do not in any way intend to indicate that we will attempt to settle that issue in this chapter. Rather we plan to make only the more limited contribution of examining the role of only one Jesuit, Melchior Inchofer, in the Galileo case in the hope that this will help to clarify one part of the larger picture.

The choice of Inchofer has been dictated by two considerations: (1) he has been largely ignored by English language scholars working on the Galileo affair, and (2) more importantly, of all the Jesuits at the time, he was the one who was most directly involved in Galileo's actual trial because of his role as an adviser of the Holy Office. In some ways he was an odd and unexpected choice for that role. But the fact is that he was selected to serve on what has come to be called the Special Commission, which twice was asked to submit evaluations of Galileo's *Dialogue,* which were then directly used in the legal proceedings against Galileo. Also either during or immediately after the trial, he wrote a lengthy assessment of the theological issues involved, which now serves as a clear barometer of his frame of mind when he participated in the trial. To what extent this treatise (Inchofer 1633) also represents the thinking of other Jesuits at the time is more difficult to determine, as we shall see. At any rate a close study of Inchofer's role in Galileo's trial will cast at least some light on the broader issue of Galileo's complaints about the Jesuits in his last years.

THE BIOGRAPHICAL BACKROUND

Since Inchofer has been generally neglected[5] by English language scholars of the Galileo affair, a few biographical remarks are in order.[6] Melchior Ildephonsus Inchofer (c. 1585–1648) was born in Hungary into a Lutheran family. Possibly because he had received his education in a Jesuit school, he converted to Catholicism. Then in March of 1607 he joined the Jesuits in Rome, and was destined to spend the remainder of his life in Italy. After his education and ordination, he was sent in 1616 to the Jesuit college in Messina, where he taught mathematics, natural philosophy, and theology for the next thirteen years.

During those years he became interested in a Sicilian legend to the effect that the Blessed Virgin Mary was the author of a locally held letter addressed to the people of Messina. The letter, dated AD 62, listed the main teachings of the Catholic religion. Inchofer became convinced of the authenticity of the letter and in 1629 published a defense of it as genuine. But as a result he was summoned by the Congregation of the Index to Rome to give an account of this book, which claimed to provide conclusive proof of the authorship of the letter. Being thus faced with the threat of his book being put on the *Index,* he changed the title and a few paragraphs to soften his claim in the book to be merely a probable conjecture and was allowed to republish it in 1631.[7]

This story is interesting as a sign of Inchofer's initial religious gullibility and submissiveness. Even of greater interest is the fact that the first objectionable edition of his book was placed on the *Index* on 19 March 1633, four weeks before Inchofer delivered his scathing denunciation of Galileo's *Dialogue* to the prosecutors at Galileo's trial, then in progress.

We thus have the oddity of an adviser of the Holy Office recommending the condemnation of someone else's book at the same point in time when his own book was being formally rejected. Moreover he was selected to be an adviser to the Holy Office for an orthodoxy assessment of Galileo's book after he himself had just been investigated by the same office. If it was politically necessary to have a Jesuit member of the Special Commission, better choices were available (e.g., Christopher Grienberger, an experienced and widely respected Jesuit astronomer and successor to Clavius at the Collegio Romano). The choice of Inchofer makes one wonder if a compromised and submissive Jesuit was wanted, or if it was simply that a theologian, not a scientist, was preferred.

At any rate Inchofer was involved in the Galileo affair for a period of only one year, from the summer of 1632 to the summer of 1633. But this was the critical period in which Galileo and his *Dialogue* were examined and condemned. After the trial Inchofer returned to Messina, and later lived again in Rome, then in Macerarta, and finally in Milan, where he died at the age of sixty-three. During these years he had nothing more to do with either Galileo or with science.

In the years after the trial he published several other books.[8] In 1639 he was chosen to give the funeral oration at the Dominican church of Santa Maria sopra Minerva in Rome for Niccolò Riccardi, O.P., who was Master of the Sacred Palace at the time when Galileo's *Dialogue* was published, and who thereby played a major role in granting the "imprimatur" for Galileo's book. So Inchofer's status was still secure in Rome after the trial.

A posthumously (1653) published booklet was his *De eunuchismo dissertatio* in which he argued on moral grounds against the practice, then in use, of castrating young boys to prepare them for later service as singers in church choirs. Oddly enough this booklet was directed against *Decisiones morales,* written in 1641 by Zaccaria Pasqualigo, who was a fellow adviser to the Holy Office for the Galileo trial. Pasqualigo's book was subsequently placed on the *Index.* Meanwhile Inchofer was strongly denounced by the church musicians. It was a precarious time for authors. In his later years Inchofer also worked on a history of the church in Hungary (only one volume, *Annales ecclesiastici regni Hungariae,* was published in 1644, which carried the history up to the year 1059). At the time of his death he was working on a book on the Roman martyrology at the Ambrosiana archives in Milan. Two other late writings were his *Tractatus syllepticus* (1633) and his *Monarchia solipsorum* (1645),which will be discussed in detail later in this and in the next chapter.

The First Report of the Special Commission

To understand how Inchofer became involved in the Galileo case it is necessary to take a look at some historical developments in 1632. In the early spring of that year Galileo's *Dialogue Concerning the Two Chief World Systems* was published in Florence and soon became widely available. Within a few months it had become a topic of considerable controversy, which took on many specific forms. Some were serious, some were trivial.

For example, did Galileo deliberately intend to insult Pope Urban VIII (whose "power of God" argument was put in the mouth of Simplicio, the simpleton, in the last pages of the *Dialogue*)? What was the true meaning of the drawing on the frontispiece of the book depicting three dolphins chasing each other in a circle? (It was merely the publisher's trademark.) Was the Preface intended to be an affront to the Roman authorities (because of its ironic wording and because it was printed separately from, and in different type than, the body of the book)? Did Galileo commit a theological offense when he argued that the certitude of human reason in mathematics was equal in strength to the certitude of the divine reason itself?

But most importantly, did the book, despite its dialogue format, defend Copernicanism? If so, it violated the condemnation of heliocentrism, which had been announced sixteen years earlier, and which would have been easily remembered by most everyone at the time. More specifically, on 5 March 1616 the Congregation of the Index issued a Decree which pronounced that Copernicanism was "false and completely contrary to the divine Scriptures," and then, after proscribing three suspect books, added that "all other books teaching the same thing are prohibited, as the present Decree prohibits, condemns, and suspends them all respectively." This left no room for doubt that any future book that defended heliocentrism would automatically be in violation of the Decree.

To make matters still worse, Galileo's *Dialogue* was published at a very disadvantageous time politically. For a number of years Urban VIII had allied himself with France, Bavaria, and Sweden against Spain and Hapsburg Germany as the Thirty Years War dragged on. This policy reached a crisis stage in the Vatican in March of 1632 when Cardinal Gaspare Borgia, who was also the Spanish ambassador, publicly charged the pope with favoring heretics and not protecting the interests of Catholic Europe. Placed in this background, Galileo's unorthodox *Dialogue* could be, and was, seen by some as a further debilitating challenge to papal authority.

At any rate the normal procedure at the time would have been to refer a book containing suspect views to the Congregation of the Holy Office for an adjudication of the matter. However for reasons that are not really known,[9] Urban VIII intervened in the ongoing controversy over Galileo's *Dialogue*. Probably in late July, he created the Special Commission, under the supervision of his nephew Cardinal Francesco Barberini, to handle the case instead of the Holy Office. There are even some hints in the documents that there was hope that it would obviate the need for a trial for Galileo. Unfortunately there are no surviving documents that establish this Special Commission and define its task. However, with one major exception,[10] scholars have concluded from somewhat indirect evidence that it seems to have had the limited specific charge of determining the facts relating to the publication of, and the required permissions for, Galileo's *Dialogue*. This evidence includes the fact that that is what is stated in the first sentence of the Special Commission's final report.[11]

By about the middle of August the Special Commission appears to have begun its work. It met five times in full secrecy, leaving no record of its internal discussions. It finished its work by writing a report that, as we shall see, made a trial for Galileo inevitable. This critically important report carries no date and no signatures. The date is less of a problem since other developments indicate that it had to have been written sometime in mid-September. The lack of signatures is unfortunate because without them the identity of the members of the Special Comission is not completely clear.

The best evidence to identify these members is contained in a letter of 11 September from Francesco Niccolini, the Tuscan ambassador in Rome, to Andrea Cioli, the Tuscan Secretary of State. Niccolini, who was concerned that the Special Commission was stacked against Galileo, reported on a meeting with Niccolò Riccardi in which the latter said,

> The Papal Theologian truly is a man of good will, and the Jesuit whom he [Riccardi] has proposed is his confidant, and he assures that he acts with good intentions. He does not know of any reason why we should be worried about them. (Galileo 1890–1909, 14:389)

The Papal Theologian at the time was Agostino Oreggio, who later was appointed a cardinal by Urban VIII. The unnamed Jesuit was almost certainly Fr. Melchior Inchofer, who was indeed a close friend of Riccardi. Riccardi himself seems to have been the third member of the Special Commission.[12] He was a logical choice in light of the Commission's role of determining how the publication permissions were secured, since he was directly involved in that process.

All of these men were primarily theologians with no background in science or astronomy, except for Inchofer who had a modest acquaintance with these areas but who was also strongly anti-Copernican in his views. Niccolini was quite concerned about this, and at the direction of the Grand Duke, he suggested to Riccardi that Thomas Campanella, O.P. (a Galileo sympathizer), and Benedetto Castelli, O.S.B. (Galileo's former assistant and an accomplished scientist), be added to the Special Commission.[13] This met with a hostile reaction from Riccardi.

> The request that Fr. Campanella and Abbot Castelli serve as lawyers and procurators could not be accepted even if the Holy Office were to wish to proceed in that way. For the former has written a similar book which has been condemned, and he could not be a defender when he himself is guilty; and the latter could not listen to the case since he is under suspicion for other reasons. (Galileo 1890–1909, 14:389)[14]

At any rate it is clear that Melchior Inchofer, who incidentally happened to be in Rome to give an account of one of his own books, entered the Galileo case because he was selected to serve as a Jesuit member of the Special Commission investigating the publication of Galileo's *Dialogue*. As mentioned earlier, he was an odd, but safe, choice because of his own immediately prior troubles with the

Holy Office. Since the Special Commission's report is unsigned, we do not know who its principal author was, although Inchofer must have had some voice in it and must have approved of its contents, however it was composed.

The report is carefully crafted as a legal document.[15] It has two unnumbered parts, each beginning at the top of a new page, and each part follows the same chronological ordering of the events described. The first and shorter part is a listing of the Commission's findings or conclusions. The second part is a more detailed listing of the facts, to which is appended as supporting evidence four additional documents, namely, copies of three letters between Riccardi and Clemente Egidi, the Inquisitor in Florence, dealing with the permissions to publish, plus a copy of the final version of the book's preface. The first part ends with this advice: "It is now necessary to consider how to proceed both against this person and in regard to the already published book." Such follow-up actions would, of course, be the proper business of the Holy Office, not the Special Commission.

In each part of the report most of the space is devoted to what happened regarding the "imprimatur," the permission to print, which Galileo had sought first in Rome and later in Florence.[16] Both imprimaturs appeared in the book, which was unusual, and which raised questions as to whether they had been properly and honestly secured. This was a serious matter because of the edicts laid down in canon law and church councils (e.g., the fourth session of the Council of Trent), which were very specific about such permissions. During Galileo's trial in the subsequent spring, much of the interrogation was devoted to this issue of the double imprimatur.

If the sole charge to the Special Commission was to establish the facts regarding these permissions, then its role was completed at this point. But the report continued on further with judgments about other matters, whether originally asked for or not. Part one of the report thus ends with the following devastating charges.

> We think that Galileo may have overstepped his instructions by asserting absolutely the earth's motion and the sun's immobility and thus deviating from hypothesis; that he may have wrongly attributed the existing ebb and flow of the sea to the nonexistent immobility of the sun and motion of the earth, which are the main things; and that he may have been deceitfully silent about an injunction given him by the Holy Office in the year 1616, whose tenor is: "that he abandon completely the above-mentioned opinion that the sun is the center of the world and the earth moves, nor henceforth

hold, teach, or defend it in any way whatever, orally or in writing; otherwise the Holy Office would start proceedings against him. He acquiesced in this injunction and promised to obey." (Galileo 1890–1909, 19:325; Finocchiaro 1989, 219)

The Special Commission had no authority to judge the case, but at most to indicate that Galileo "may have" been guilty of several offenses. First he "may have" treated the motion of the earth and the stability of the sun absolutely rather than hypothetically. This is almost certainly a reference to Bellarmine's Letter to Foscarini of 12 April 1615, which makes precisely the same distinction, and which subsequently became standard parlance in Rome regarding Copernicanism.[17]

It is important to note that at that time "hypothetical" did not have its contemporary sense of a plausibly true claim about the world for which more data is to be sought for verification. Rather it meant a counterfactual claim, that is, one taken to be not true of the world, but which, if arbitrarily assumed, is convenient for calculational purposes (Blackwell 1991, 80). And "absolute" here meant "really true of the actual world." Bellarmine had no objections to such a "hypothetical" use of the Copernican theory, but rejected it in its "absolute" sense as contrary to the Scriptures and the religious tradition.

The second charge applies this same distinction to Galileo's theory of the tides in the *Dialogue*. Although this Galilean theory turned out to be notoriously false on empirical grounds, that is not the complaint made in the report. Rather the point was that if the *actual* facts about the tides are explained by Galileo's double motion of the earth, then the latter as their *actual* cause is taken absolutely, not hypothetically.

But the real bombshell was the last point. To almost everyone's surprise, including even the pope, the Special Commission had somehow uncovered a memo in the files of the Holy Office,[18] part of which it quotes verbatim, regarding Galileo's meeting with Bellarmine on 26 February 1616. This memo says that an injunction was delivered at that time to Galileo not to "hold, teach, or defend it [Copernicanism] in any way whatever, orally or in writing," and that he had agreed to obey this order.[19] As a result of this new information Galileo's *Dialogue* was then to be judged not simply against the general condemnation of Copernicanism in the Decree of 5 March 1616 but also against this much more restrictive personal injunction. And it seemed indeed highly likely that the *Dialogue* would have violated this much stronger order. As a result a trial for Galileo before the Holy Office had become inevitable.

The devastating nature of this development can be seen in Riccardi's comment as reported by Niccolini.

> But above all he [Riccardi] said, under the usual conditions of confidence and secrecy, that it has been discovered in the records of the Holy Office that, about twelve years ago when it was learned that he held this view [Copernicanism] and was spreading it in Florence, Galileo had been called to Rome where, in the name of the Pope and of the Holy Office, he was prohibited by Cardinal Bellarmine from holding that view. This alone is enough to utterly ruin him. (Galileo 1890–1909, 14:389)

Sometime during the late summer of 1632 Pope Urban VIII's attitude toward Galileo changed dramatically from one of friendship to one of anger and rage against him.[20] He said repeatedly to Niccolini that Galileo had deceived him about this injunction, which previously had been unknown to him. He also concluded that a trial was inevitable. The report of the Special Commission was presented on 23 September to the Congregation of the Holy Office with the pope in attendance. The meeting concluded with the pope's order that Galileo come to Rome in October to answer the charges against him.

Although Melchior Inchofer's name does not explicitly surface in the documents relating to these events, we know that he was directly involved, and that through his work on the Special Commission he played a role in uncovering, and bringing attention to, the 1616 injunction against Galileo, which would "utterly ruin him." He in effect was the co-author of the indictment. Galileo's suspicions that the Jesuits had a hand in his downfall were justified in at least this case.

THE SECOND REPORT OF THE SPECIAL COMMISSION

Because of a series of delaying tactics, Galileo did not arrive in Rome until February of 1633, and the first interrogation at his trial did not occur until 12 April. A short time before that date the Special Commission was reconstituted, and then called upon a second time for an opinion in the case.[21] The specific issue up for analysis this time was whether or not the *Dialogue* violated the conditions of the injunction of 1616, namely, that Galileo was "not to hold, teach, or defend it [Copernicanism] in any way whatever, orally or in writing." These opinions were submitted on 17 April and were later used by the court in deciding the case.[22] Unlike the Special Commission's evaluations in the previous September, this time each member delivered his own separate and personally signed report.

Judging from the structure of these reports, Copernicanism was defined for the commission in terms of two distinct propositions: (1) the sun is at rest at the center of the universe, and (2) the earth revolves around the sun. The question to the commission, then, was whether Galileo in the *Dialogue* had "taught, defended, or held" either or both of these claims. If so, he stood in violation of the injunction.

This request for separate evaluations of the sun's rest and the earth's motion, which may seem odd to the contemporary reader, was made for theological, not scientific, reasons. In terms of the astronomy of that day, (1) implied (2), and vice versa, and thus both propositions stand or fall together.[23] But in regard to scriptural exegesis, there was an important difference here because various passages in the Bible seem to *directly* contradict (1), which would thereby be "formally heretical," while (2) is only *indirectly* contradicted, and thus would fall into the less serious category of being "erroneous in faith." Exactly this same appeal to these two levels of evaluation had also been used in February 1616, when the Holy Office asked its consultors at that time for their opinion on the orthodoxy of Copernicanism.[24] This indicates that the Holy Office was using that earlier file in setting up Galileo's trial in 1633.

Of the three reports (Galileo 1890–1909, 19:348–60; Finocchiaro 1989, 262–76) submitted to the court, Inchofer's was by far the longest and the most hostile to Galileo. Its first part deals with the immobility of the sun at the center of the universe, and the second, longer part discusses the earth's motion. In both cases Inchofer concludes that Galileo in the *Dialogue* "teaches, defends, and holds" that view, and thus is guilty of violating the injunction across the board.

In making his case Inchofer carefully defines each relevant term in the injunction. He quotes Augustine to establish the notion that "to teach" means simply to communicate knowledge to someone else.[25] He then has no difficulty showing that the *Dialogue* does this, since it clearly transmits and explains the meaning of the Copernican theory to its readers. Even if Galileo's purpose had been to refute heliocentrism, he would have "taught" it in this sense.

Next he defines "to defend" as meaning to attack the arguments supporting the opposite position. Again, given this definition, Galileo's book clearly "defends" Copernicanism, especially because its dialogue format allows him to attack relentlessy and with vigor the evidence and arguments adduced in favor of the Aristotelian geocentric world view, which was supported by Simplicio.

It is important to note that in this part of his report Inchofer also explicitly mentions by name his fellow Jesuit, Christopher Scheiner, a recent opponent of Copernicanism, who he says is the principal target attacked by Galileo.[26] The particular point at issue was Galileo's argument for heliocentrism from the phenomenon of the variation of the curved paths of sunspots. This appears in the middle of the third day of the *Dialogue,* where Galileo rejects the ideas of Apelles (Scheiner's pseudonym) as "vain and foolish." It is well known that Galileo and Scheiner had been engaged in a bitter dispute for the previous twenty years over the discovery and interpretation of sunspots. Could it be that Scheiner was actively advising Inchofer in his reading and evaluation of the *Dialogue* as he was preparing his report?

The third issue of whether Galileo "holds" heliocentrism was a more difficult question. The reason is that "to hold" was taken to mean that one personally believes that something is really true of the actual physical world. This obviously would involve a claim about Galileo's subjective state of belief, which was not publicly accessible, unlike the cases of "to teach" and "to defend." As Inchofer himself puts it, this is a "question of Galileo's mental attitude."

As a result one must rely on indirect evidence to judge this part of the injunction. What follows is a complex discussion in which Inchofer gives six arguments and twenty-seven textual quotations from the *Dialogue* to justify his conclusion that Galileo is "vehemently suspected of firmly adhering to it [Copernicanism], and indeed that he holds it." It is interesting to note that at the end of the trial, Galileo was judged to be "vehemently suspected of heresy" [this was a technical legal expression used by the Inquisition], where "suspected" is weaker than Inchofer's "indeed . . . he holds it."

Inchofer's arguments on this point are based on the principle that "the mind of a speaker is tied to his words," and hence one can reason from the latter to the former. But the *Dialogue* often presents Copernicanism as the true account of the world. And why did Galileo not defend the 1616 Decree against Copernicanism? Why did he write in Italian, if not to "entice that view in common people?" Why did he praise William Gilbert, "a perverse heretic"? In an apparent reference to Galileo's *Letter to the Grand Duchess Christina* (written in 1615—long before the present set of complaints against Galileo and prior to the Decree of 1616), Inchofer claimed that Galileo "not only proved Copernicus's opinion, but established it by solving scriptural difficulties as well as he could." The long list of passages quoted from the *Dialogue* is focused on showing, from Galileo's own words, that in his mind he personally held a belief in heliocentrism.

Inchofer's main argument on this point, however, hinges on the "absolute" vs. "hypothetical" distinction. For him Galileo initially claims to argue by means of a mathematical hypothesis, which by its very nature is merely a supposition not requiring any physical proof and not having any physical consequences. But Galileo later goes beyond this by attempting to prove that his hypothesis is actually true of the physical world. For Inchofer this is a violation of the distinction between mathematics and philosophy. The former is purely hypothetical, while only the latter can draw existential conclusions about the actual world. This is a view about the relationship between these two disciplines that in the recent past had come to be strongly advocated at the Jesuit college in Rome.[27]

As a result Galileo is depicted as having come to believe personally in Copernicanism by using a methodology that entailed an illicit transition from the hypothetical to the absolute. A mathematicized science that also arrives at physically true conclusions was a concept foreign to Inchofer, although it was one of Galileo's grand insights. But for Inchofer astronomy cannot draw physical conclusions, only make calculations based on convenient assumptions. As he put it:

> So, in order to restrict himself to a pure mathematical hypothesis, Galileo did not have to prove absolutely that the earth moves, but only to conceive its motion in the imagination without assuming it physically, and thereby explain celestial phenomena and derive the numerical details of the various motions. (Finocchiaro 1989, 267)

Whether Inchofer's argument here really proves that Galileo "holds" Copernicanism, in the sense indicated, is indeed very questionable. But what is more interesting is what this shows about the lack of understanding, on the part of Galileo's ecclesiastical critics, of the epistemological character of the new science that was emerging in his writings. Scheiner could have, and indeed should have, given him better advice on this. Be that as it may, Inchofer's final opinion in his report to the court is that Galileo teaches, defends, and holds both the earth's motion and the sun's immobility, thereby putting him in violation of the injunction. The court would later adopt this same judgment.

In summary Inchofer had been asked merely to give his considered opinion about the relation of Galileo's *Dialogue* to the injunction. It is notable that he went beyond this in his rhetoric to submit an unusually harsh assessment of Galileo and his writings.[28] One wonders whether the tone of sharpness, and perhaps even of personal hostility, which appeared only in his report, indicates that he may have been speaking for some other Jesuits as well, particularly Scheiner. But I have found no evidence to prove that assertion.

The reports from the other two Commission members came basically to the same final conclusions as did Inchofer. Oreggi, the senior theologian, submitted only one paragraph to this effect. He gave no specific reasons or textual quotations to support his judgment, saying only that the whole thrust of the *Dialogue,* plus the content of the Special Commission's first report, was enough to justify his opinion.

In many ways the most thoughtful and balanced report was submitted by Pasqualigo, who at the age of thirty was the youngest member of the Commission. He also maintained that Galileo argues from the hypothetical to the absolute. But he supported this with a surprisingly well-informed discussion of technical astronomical arguments in the *Dialogue.* After recalling the wording of the injunction, Pasqualigo concluded that Galileo does teach and defend Copernicanism, and that he is "strongly suspected of holding such an opinion" (but unlike Inchofer, he does not go beyond that point), and therefore he has transgressed the injunction.

Most modern readers of the *Dialogue* would probably come to approximately the same conclusions about what Galileo had presented in his book as did the members of the Commission. Galileo did indeed clearly seem to advocate Copernicanism in the senses indicated. But today's readers would undoubtedly focus instead on the wisdom and the legality of the injunction itself, which was at the center of the case being assembled against Galileo. But of course the three members of the Commission would not, and could not, have ever considered making that suggestion, even if it had occurred to them to do so.

With the submission of his report to the court, Inchofer's role in the trial came to an end. But his attack on Galileo and Copernicanism was just beginning. For behind the scenes, while Galileo's trial was still underway or immediately thereafter, Inchofer wrote a ninety-three-page book entitled *Tractatus syllepticus.*[29] His aim as stated in the sub-title was to identify what must be accepted as the truth about the motion and rest of the sun and the earth, as this is determined by the scriptures and the church fathers. The book was published in Rome in the fall of 1633, shortly after the trial ended.

It is obvious from several indications that Inchofer was not motivated by a theoretical interest in this question. Rather he had in mind the polemical and political goals of closing the Copernicanism question once and for all on theological grounds, and in the process providing a defense of the trial judgment against Galileo. Also Inchofer's now largely neglected book, which deals with the central issue of the relations between natural science and scripture, which was at the heart of Galileo's trial, provides a unique and direct insight into the theological thought operating behind the trial records. (The whole of chapter 3 will focus on this issue.) In effect the *Tractatus syllepticus* was the first in a long series of ecclesiastical apologetic reactions in the years after the trial, which were intended to defend the Church against strong negative reactions, which were already anticipated.

INCHOFER'S DOWNFALL

Inchofer's personal life and fortunes began to change dramatically after 1629. As we have seen, he was able to negotiate a rather simple modification of his book on the letter of the Blessed Virgin Mary when challenged by the Congregation of the Index. This in turn evolved in time into a strong and close friendship with Fr. Riccardi, which lasted until the latter's death in 1639, with Inchofer invited to deliver the funeral oration. Possibly through Riccardi he became well acquainted with Cardinal Francesco Barberini, the pope's nephew, who consulted with him on several occasions into the 1640s. Inchofer seemed to be attracted to such centers of power, and he knew how to keep in their favor.

On the other hand, during these same years he began to fall out with his fellow Jesuits in ways that ultimately became self-destructive. This started with his two writings against Copernicanism: the *Tractatus syllepticus* and another unpublished manuscript called the *Vindiciarum S. Apostolicae*. Some of the Jesuit referees of these books had objections to both of them, especially the latter, which was blocked from publication, because they thought that Inchofer was naively overstating the case against heliocentrism from the appeal to sacred scripture (which in turn implies, by the way, that the Jesuits as a whole were not involved in a conspiracy against Galileo). Not being open to compromises with the Jesuits, Inchofer gradually became more and more scandalous and even defaming, if not libelous, about various Jesuits and their customs and history.

Behind all this was also the fact that Inchofer had a quite odd personality: he could be very difficult to work with; he had some peculiar interests and convictions; and he was not predictable but rather "marched to his own drummer," as the saying goes today. To use the adjectives of Thomas Cerbu (2001), he was sly, crafty, and quirky. For example, he published books sometimes under his own name, sometimes under various pseudonyms, sometimes even under someone else's name, and sometimes anonymously. His Jesuit confreres had reasons to be wary of him. His allegiance was shifting from his fellow Jesuits to the decision makers at the Vatican.

This all came to a climax during the last year of Inchofer's life. In 1645 an anonymous volume entitled *Monarchia solipsorum* was published in Venice. This book was such an offensive defamation of the Jesuits that they tried to buy up all outstanding copies, often at highly inflated prices. The question, of course, was who was the author, and soon Inchofer was mentioned as a possible culprit. In time a search of his personal papers revealed that he had sent slanderous materials to other critics of the Jesuits, and that he himself had written in his own hand further chapters to be added to the *Monarchia solipsorum*.

In January 1648 Inchofer was put on trial by the Society of Jesus under a provision advocated by a well known Jesuit legal scholar, Francesco Amico, that a religious order can formally prosecute a disloyal member for defamation of the community. The trial lasted ten days and ended when Inchofer confessed to the charges against him. He was sentenced to one month of penance, to a loss of his voice in the affairs of the Society, and to an indefinite term of imprisonment at the discretion of the General of the Society. Inchofer died later that year on 28 September.

This story could hardly be more ironic. Galileo's foremost critic at his trial, who was the sharpest and most damaging in his denunciation of the *Dialogue*, was fated himself fifteen years later to undergo a trial, condemnation, and sentence which had eerie similarities to what had happened to Galileo.

The Scriptural Case against
Copernicanism in 1633

THE DOCUMENTS DETAILING THE PROCEEDINGS AT GALILEO'S TRIAL OF
1633 do not present scriptural or theological arguments to establish the point
that Copernicanism is a religiously unorthodox point of view. That was simply
presumed to be a settled matter. The issue rather was only whether Galileo's
Dialogue advocated the heliocentric position in violation of the Congregation
of the Index's Decree of 5 March 1616 and of the disputed injunction served
on Galileo a few days earlier. For the justification of the Decree and the injunction one must go back to the scriptural and theological discussions that occurred
seventeen years before the trial.

Had the assessment of Copernicanism evolved or been changed in any
way during those years? Fortunately we are at least able to entertain that question because one of the key participants at the trial wrote a lengthy treatise,
largely ignored by Galileo scholars, which reexamined in detail the scriptural
and theological case against heliocentrism. This tract was written while the trial
was being conducted, and thus in a special way it gives a direct insight into at
least one participant's view of the religious rationale behind the trial. The treatise is entitled *Tractatus syllepticus,* and was written by Melchior Inchofer, S.J.,

who earlier had submitted an assessment requested of him by the court concerning the orthodoxy of Galileo's *Dialogue.*

The Context of the *Tractatus*

Inchofer's book, which he conceived to be the first of two treatises on the orthodoxy of Copernicanism, was published in 1633. He probably completed his composition of the book sometime in late July, just a few weeks after the end of Galileo's trial, since the permissions to publish are dated to mid-August. The bulk of the book was likely written during the months of Galileo's trial. We do not know exactly when Inchofer began work on this project, but M. J. Gorman (1996, 291) claims that, "The work seems to have been begun before February 13, 1633." At any rate Inchofer's writing of the *Tractatus* was virtually simultaneous with the trial itself.

Also at the same time another Jesuit, Christopher Scheiner, wrote a treatise entitled *Prodromus pro sole mobile* (1651), attempting to refute Copernicanism. At the time, Scheiner was a world class observational researcher and authority on solar astronomy. It thus appears that Inchofer and Scheiner together constituted a double-barreled attack on Galileo's *Dialogue,* on the theological and scientific levels respectively. For some complex reasons that we will examine in chapter 4, Scheiner's *Prodromus* was withheld from publication by the Jesuit censors and was not published until 1651, one year after his death.

Inchofer's *Tractatus,* however, was granted permission for publication by all three of the censors assigned to evaluate it, and was accordingly approved by the General of the Jesuits at the time, Muzio Vitelleschi, on 18 August (see appendix 1.). However Vitelleschi does not mention that one of the three censors, Christopher Scheiner, added a follow-up note to his approval, which reads as follows.

> It appears that the author asserts too absolutely at the beginning of page 34 that the motion of the sun and the immobility of the earth are matters of faith, which should be modified since they are in question, and are not thought to be true matters of faith. Moreover he should also indicate briefly for what reason this might be a matter of faith. I believe that similar considerations ought to be taken into account with respect to the circular motion

of the sun and the center of the earth being in the middle of the universe. (Gorman 1996, 295–96; "page 34" in the quotation refers to Inchofer's autograph, not the printed, text)

As we shall see later, Scheiner had placed his finger on the central and most contentious claim in the *Tractatus*.

As we have mentioned earlier, Inchofer had prepared a second treatise, obliquely referred to several times in the *Tractatus,* which continued the attack on Copernicanism. Its full title was *Vindiciarum S. Sedis Apostolicae, Sacrorum Tribunalium et Authoritatum adversus Neopythagoraeos Terrae motores, et Solis statores, libri duo.*[1] The censors rejected this book, which then was never published. The reasons were that (1) the title presumptuously implies that the author speaks for the Holy See (Inchofer's later offer to avoid this by changing the title did not work this time), and (2) the arguments are not sufficiently strong, powerful, or solid (see appendix 2).

In the light of these internal disputes among the Jesuits most directly involved in the Galileo trial and its immediate aftermath, it is worth noting that the Jesuits did not display a united front to oppose Galileo, but rather were in disagreement among themselves. It has even been suggested by some that Scheiner was a "closet Copernican" (e.g., Gorman, 1996, 307).

In reply to his critics Inchofer offered two justifications for his views in the *Tractatus*. First, Galileo had been condemned and forced to reject the doctrine of Copernicanism, which alone is an adequate reason to say that heliocentrism is heretical. Second, Inchofer said he never would have dared to say that without having first consulted with those who have the legal authority to decide such matters.[2] On the first point it is interesting to see that the Galileo trial had immediately created a need to examine in theological detail the notion that heliocentrism is formally heretical. That is the project Inchofer undertakes in the *Tractatus*. On the second point Inchofer seems to quite clearly imply that the *Tractatus* was written with the approval, or with the encouragement, or perhaps even in response to the direct or indirect request of Urban VIII or one of his representatives. If so, this treatise takes on a much more official status than first meets the eye.

This interpretation is supported by an examination of the iconography on the title page of the *Tractatus* (see fig. 1). The outer limit of the drawing is an equilateral triangle, which represents the triune God of Christianity. Inscribed in the triangle is the earth, which is held in place by three bees, one in each

MELCHIOR(IS) INCHOFER

E. SOCIETATE IESV

AVSTRIACI,

TRACTATVS SYLLEPTICVS,

In quo,

QVID DE TERRAE, SOLISQ. MOTV, VEL STATIONE,
secundùm S. Scripturam, & Sanctos Patres sentiendum,
quauè certitudine alterutra sententia tenenda sit,
breuiter oftenditur.

ROMÆ,
Excudebat Ludouicus Grignanus MDCXXXIII.
SVPERIORVM PERMISSV.

FIGURE ONE. Title page of Melchior Inchofer's *Tractatus syllepticus* (1633). Reproduced with permission from the Special Collections Library, University of Michigan.

angle of the triangle, whose legs are firmly implanted in the triangular God, and whose front legs and antennae grasp the earth. These bees, an unmistakable reference at the time to the Barberini family symbols of Urban VIII, serve as the conduits connecting God to the earth, which indeed is the role of the pope. Lest we still are unaware of the main point, above the icon are inscribed the words "HIS FIXA QUIESCAT" (It is at rest, held firmly by these bees). Or in theological language, the pope has firmly established the doctrine that the earth is at rest. Heliocentrism is heretical.

This icon on the title page is not as original as one might at first think. The most widely known event in Rome in June 1633 was not the conclusion of Galileo's trial on the 22nd but the consecration, exactly one week later on the 29th, of Bernini's long awaited baldaquin in St. Peter's Basilica. The baldaquin is the famous gigantic canopy held up over the main altar by four large twisted support columns. On the very top of the canopy is a huge globe of the earth from which protrudes at the north pole a large cross, which together symbolize the Christianization of the earth. If one looks closely (the view from directly below the canopy is best), one sees that the earth at the top of the baldaquin is held up by a swarm of bees. No one in Rome at the time would have missed the point.

In placing this imagery on the title page of the *Tractatus* was Inchofer boldly using papal authority to his advantage, or did he have some sort of official approval or encouragement? The latter seems much more likely.[3]

Shortly after the publication of the *Tractatus* several of Galileo's friends informed him of it by letter, complaining that Inchofer was creating new articles of faith. Galileo's most explicit comment about Inchofer's book is contained in a letter he wrote, while under house arrest at Arcetri, to Elia Diodati (25 July 1634).

> Recently a Jesuit priest has published a book in Rome, which claims that this view [the motion of the earth] is so horrible, pernicious, and scandalous that, if it were allowed in the schools, in group discussions, in public disputes, or in books, it would threaten the main articles of faith; namely, the immortality of the soul, creation, the incarnation, etc. Therefore no one should be allowed to debate or to argue against the stability of the earth. This article alone, above all others, should be held in such high protection that no one should ever entertain the contrary, either for purposes of debate or even to give it stronger support. The title of this book is *Tractatus syllepticus* by Melchior Inchofer, S.J. (Galileo 1890–1909, 16:118)

Thus less than one year after the publication of Inchofer's book, Scheiner's original concern about creating new articles of faith had begun to spread. Was there something contained in Inchofer's basic conception of theological truth that opened the door to such a development? In attempting to give a theological justification for the verdict in Galileo's trial, had Inchofer modified or elaborated in some way the Church's reasoning that stood behind the Decree condemning Copernicanism in 1616? Perhaps a close conceptual analysis of the *Tractatus* will tell us.

Delineating the Goal of the *Tractatus*

Inchofer's frame of mind as he takes up his pen is directly stated in the first two sentences of his preface. Copernicanism has been, and ought to have been, forbidden by the Church as false, a clear reference to the Decree of the Congregation of the Index of 5 March 1616. Nevertheless many are still "itching" to speak and write in its defense, and others are "itching" to hear this, as Luke Wadding, an Irish Franciscan official in the Curia, puts it in his prefatory assessment of Inchofer's book. He even goes on to claim that this treatise proves that all the "human sciences should be subordinated to the rules of Sacred Scripture," a much wider claim than Inchofer would need. To put an end to this abuse, Inchofer will focus on the theological rationale behind the condemnation of Copernicanism and Galileo, leaving the legal rationale to the recently pronounced verdict of the inquisitional court and the scientific justification to those versed in mathematics, that is, to Christopher Scheiner, who was simultaneously writing his *Prodromus*. In short, for Inchofer to provide the Church with a theological justification of the verdict in the Galileo case, he must show that the Decree of 5 March 1616 is well grounded in the sacred scriptures, which serve as the basis of theological truth.

In the first few short chapters of the *Tractatus* Inchofer locates the cause of the "itching" in the ambiguity that one can easily find in the scriptures. For there are many passages that seem to say that the sun moves and that the earth is at rest, and many other passages that seem to say that the sun is at rest and the earth moves. Overcoming these ambiguities requires a distinction between meaning and truth, to be followed by an elaborate taxonomy of levels of meaning, and then by a determination of the degree or strength of truth claimed for the theological rejection of the heliocentric point of view.

It is important for what follows in the *Tractatus* to see that truth presupposes meaning, but not vice versa. The statement "All As are Bs" is neither true nor false because it says nothing, it has no meaning beyond its own symbols. If we change it to be "All planets revolve around the sun," it now has meaning, but its truth remains indeterminate. Whether that statement is taken to be either true or false depends on what criterion of truth determination one accepts: perhaps on what further evidence may indicate, on the acceptance of the authority of another person, on divine revelation, on cultural conditioning, and so forth; in short, on whatever is taken to be the proper criterion of truth in a given case. Of course, in the concrete development of human knowledge, meaning and truth reciprocally influence each other, as we shall see, but the initial priority of meaning over truth is the natural starting point for a treatise on scriptural exegesis like the *Tractatus*. As a result Inchofer's first task, after delineating the problem of ambiguity on what the scriptures say about the heavens, is to distinguish the different types of meaning that can be discovered in the Bible.

To use just the first of Inchofer's examples: Psalm 97:4 says, "His lightning lights up the world, which the earth sees and is moved." Surely this passage, which seems to say that the earth is moved (and "sees"), does not really mean that any lightning bolt was so strong that it caused the earth to move. So Inchofer says that the word "earth" here really refers to the "inhabitants of the earth" who were frightened by the lightning. That is the real "literal" meaning, that is, the meaning of the words.

THE TAXONOMY OF SCRIPTURAL MEANINGS

In undertaking his discussion of types of meaning in Scripture, Inchofer of course is the beneficiary of a long exegetical tradition going back to the first centuries of the church. As a result, although his distinctions already have a long history, which he would intend to follow, his precise designations and emphases on how the various categories of meaning are to be understood, and some of the consequences that follow from these designations, introduce some distinctive characteristics in his interpretations of scripture, as we shall see.

He begins with the presupposition that the reader is already quite familiar with Aquinas's discussion of the topic in his *Summa theologiae,* part I, q.1, art. 10, which he quotes in the first sentence of chapter 5. The general framework from

Aquinas is that overall there are two levels of meaning in scripture. The first basic level is what is called the "literal or historical meaning," which refers to the things that the author intends to signify by the words used. This signification may be either direct or indirect because the author may in the latter case use some kind of a figure of speech. Thus the "arm of Abraham" has a direct literal meaning, while the "arm of God" is an indirect signification of God's power. Since the contemporary reader is not accustomed to this double sense of "literal meaning," in our discussion we will refer to these two components of literal meaning as the *factual literal* meaning and the *figurative literal* meaning. All words used in scripture have one or the other type of literal meaning. Otherwise they would refer to nothing, and would be merely nonsense sounds or gibberish. Also the literal meaning is what the author intends, a point from Aquinas upon which Inchofer places great emphasis.

The second type of scriptural meaning is called the spiritual or mystical meaning. This occurs when the things referred to in the literal meaning themselves also signify something else. Thus the word "manna" literally means the food that fell from the heavens to feed the Israelites lost in the desert, but also spiritually it refers in foresight to the eucharistic bread consecrated by Jesus at the last supper. From this notion of spiritual meaning as what is signified by what was previously signified by the words used, it follows that literal meaning must always precede spiritual meaning. Although it is not particularly relevant to the case of Copernicanism, Inchofer puts strong emphasis on the point that the spiritual meaning must be based on a prior literal meaning, for otherwise it would be subject to the vagaries of human choice and imagination uncontrolled by the words of the Holy Spirit in the scriptures.

Furthermore many, if not most, literal meanings do not have any spiritual meanings. The subtypes of spiritual meaning recognized by Aquinas are three-fold: the allegorical sense (items in the Old Law that signify things in the New Law), the moral sense (Christ's words and deeds that signify what we ought to do), and the anagogical sense (things that signify eternal life.) At various places in his text Inchofer mentions the spiritual or mystical senses of the Scripture, but he does not develop that level of meaning in any detail. The reason is that questions of spiritual meaning do not arise in any significant way in the scriptural assessment of the Copernican hypothesis.

The issue at hand then in the *Tractatus* is whether the passages of the Bible that speak of the motion of the sun and the immobility of the earth are to be understood properly as being factually literal or figuratively literal in their meaning. A very large percentage of the pages in the *Tractatus* are devoted to a text by text evaluation of this topic. Hence if Inchofer can show that the relevant words in the scriptures regarding the structure of the world support geocentrism factually and not just figuratively, and if the standard of truth applicable to the Bible is accepted, then the case against Galileo and Copernicanism will have been made. So far up to this point there is nothing new or unusual in the way that Inchofer has set up the problem of theologically evaluating heliocentrism.

The issue at hand is complicated by a further factor, namely, that the words used in scripture often have more than one literal meaning. One of the clearest examples of this stands out prominently in the present dispute: "The earth stands still forever" (Ecclesiastes 1:4). Is the literal meaning that the earth does not move from one place to another, that it does not rotate on its axis, that it does not change its relation to other bodies, that it is eternal in its past existence or in its future existence, or some combination of these? The exegete needs to sort out this plurality of literal meanings and evaluate them in terms of other scriptural passages, the history of the opinions of earlier exegetes, the common agreement of the church fathers, the words used in the original languages, and any other interpretive tools available. One of the reasons for the considerable length of Inchofer's book is the need to discuss in detail so many passages which have multiple literal meanings (which he does especially in chapter 13).

We might add that when Inchofer says that the literal meaning is the "intent of the author," the latter of course is the Holy Spirit, and not the human scribe of the autograph text. The role of this scribe is not discussed, and thus the scribe was most likely considered by Inchofer to be a neutral conduit between God and the written revealed text. Also he understands meaning to be fully objective, in other words, it is not affected by the cultural perspective of the listener. Inchofer, of course, would not deny such differences of personal perspective (otherwise there would be no controversy over Copernicanism to begin with), but they are not relevant for him to the meaning of the scriptures. How the latter is determined, and by whom, is yet to be discussed

THE TAXONOMY OF RELIGIOUS TRUTHS

Discussion in any amount of detail of the meaning of scriptural passages does not in itself establish the truth status of the results produced. One also needs to add that the established meaning is true because of the veracity of the author of the text. In other words the scriptural passages are true simply and only because that is what God has said. Elsewhere (Blackwell 1991, 102–9) I have called this *de dicto* truth ("because that is what was said"), as compared to the common sense notion of *de facto* truth ("because that is what actually happened"). The former, not the latter, is the basic criterion of scriptural truth (whether these two types of truth conditions agree or disagree in a given case, assuming that the meaning has first been correctly established.) If one does not accept the notion of the *de dicto* truth standard, then the truth status of religious claims remains indeterminate. For the main point is that the author of scripture is, of course, taken to be the Holy Spirit, who is Truth itself. In short the "author" is the ultimate "authority" on truth for the religious believer. The whole account includes the willingness on the part of the believer to accept that authority as the guarantor of truth.

This is very similar to our everyday common sense beliefs. Much of what we know—from newspapers, television news, conversations with friends and family, and so on—we believe because we are willing to accept the credibility of the source of the information. We may have been deceived, deliberately or not, or we may have misunderstood, but natural belief involves a decision to accept the authority of the source of our information as reliable. So does religious belief.

Furthermore another key characteristic of the Catholic religious tradition is the question of who undertakes the task of interpreting the meaning and truth value of the scriptures. This was a central issue of dispute during the Reformation, which resulted in the following policy of the Catholic Church formulated at the fourth session of the Council of Trent (8 April 1546):

> Furthermore to control petulant spirits, the Council decrees that in matters of faith and morals pertaining to the edification of Christian doctrine, no one, relying on his own judgment and distorting the Sacred Scriptures according to his own conceptions, shall dare to interpret them contrary to that sense which Holy Mother Church, to whom it belongs to judge of their true sense and meaning, has held and does hold, or even contrary to the unanimous agreement of the Fathers, even though such interpretations shall never at any time be published. Those who do otherwise shall be identified by the ordinaries and punished in accordance with the penalties prescribed by the law. (Blackwell 1991, 11–12)

This regulation clearly limits the interpretation of Scripture to the bishops of the church and is the basis for the system of censoring books on theology, which was established later at the Council of Trent. Notice also that this policy is explicitly focused on "matters of faith and morals" (*in rebus fidei et morum*), which are the essential matters of belief that all Catholics are expected to accept to be members of the Church. Unanimous agreement among the fathers of the church is stated as one of the conditions to judge the orthodoxy of interpretation. Other exegetical standards are also presupposed, such as internal consistency, comparison to other passages in Scripture, tradition, specific words used, and so forth. All of these factors enter into Inchofer's methodology in the *Tractatus*.

At any rate a key feature of scripturally based belief is that the source or author is always taken antecedently to be reliable. If so, then virtually all of our attention and thought processes are on the levels of factual vs. figurative literal meaning in what is revealed, which we have examined in the previous section of this chapter. Also as a result Inchofer's treatise focuses almost completely on the questions of the meaning of the biblical comments about the structure of the universe

This is unfortunate because in the process he introduces some subtle modifications concerning the truth status of what he wishes to conclude. To understand the shift that occurs here, it is necessary to review an incident that had a major impact on the formulation of the Decree of 1616 against Copernicanism.

Early in 1615 a small booklet entitled *Letter on the Motion of the Earth* was published in Naples by a Carmelite priest named Paolo Antonio Foscarini.[4] The author does not claim that Copernicanism has been proven to be the true system of the world, but only that if this were to be proven, then the Church would need to reinterpret the passages of scripture that seem to say the opposite. Foscarini's booklet was to be the first tentative step in that direction. An anonymous censor, however, objected to the book as being "rash," which is one step short of "light suspicion of heresy." In self-defense Foscarini appealed the case to Cardinal Bellarmine, who responded with what has become the famous Letter to Foscarini (12 April 1615) since it became the foundational guide to the church's assessment of Copernicanism.[5] One of Foscarini's defenses was that the issue of the structure and motions of the solar system is not a matter of faith and morals, and so is not subject to censure under the provisions of the fourth session of the Council of Trent. The key part of the response by Bellarmine to this point reads:

> Nor can one reply that this is not a matter of faith, because even if it is not a matter of faith because of its subject matter [*ex parte objecti*], it is still

a matter of faith because of the speaker [*ex parte dicentis*]. Thus anyone who would say that Abraham did not have two sons and Jacob twelve would be just as much of a heretic as someone who would say that Christ was not born of a virgin, for the Holy Spirit has said both of these things through the mouths of the prophets and the Apostles. (Blackwell 1991, 266)

We are now at the heart of the matter. Bellarmine has significantly shifted, and thereby expanded, the notion of "matters of faith and morals" from the character of the subject matter stated in the revelation (traditionally what one has to believe for salvation) to the power of truth possessed by the Holy Spirit as the speaker of the revelation. This has the consequence of granting absolute truth to whatever is said in the scripture, assuming of course that the correct literal meaning has first been established, no matter how major or minor the point might be as far as its religious impact is concerned. In effect, the criterion of truth in the scripture, the authority of its author, has converted all correctly understood information into absolutely certain beliefs that are "matters of faith and morals." Anything contrary to such beliefs would then be heretical. It is not clear that Bellarmine himself would have carried his distinction of *de fide ex parte objecti* and *de fide ex parte dicentis* this far. But Inchofer does.[6] And it enables him to formulate a theological justification for the condemnation of Galileo and Copernicanism.

As Inchofer develops his case in the *Tractatus,* his categories of *de fide* truth become more elaborated.[7] Matters of faith are revealed in numerous ways: (1) either (a) *directly* or *primarily* (explicitly stated), or (b) *indirectly* or *secondarily* (by inference that the opposite is contrary to the faith; (2) either (a) *immediately* (not contained in another revealed proposition) or (b) *mediately* (contained in another revealed proposition); and (3) where (2b) can be either *total* or *partial* containment.

The possible combinations of these categories of *de fide* truth result in sub-categories of varying strength, which are not altogether clear. But the more basic question is whether *everything* said in the Bible, *no matter how trivial,* becomes thereby a *de fide* truth, assuming of course that the correct literal meaning has first been established. Bellarmine had said that Abraham's having two sons and Jacob twelve was of equal *de fide* status as the virgin birth of Christ. He had said that only in a letter, of course, but it was a letter responding to an appeal of a censure, which then later came to partially define the Church's view after 1615.

It turns out that Galileo and his friends had correctly identified the *Tractatus* as multiplying *de fide* truths quite unnecessarily. One of Galileo's examples of this (in his unpublished notes of course) appeals to the book of Tobit in the Old Testament where Tobias is described as having an unnamed dog as his companion. Is then the claim that Tobias had a dog a *de fide* truth? Yes, Galileo mocks, because of the mere fact that that is what the Holy Spirit has said.

THE CASE AGAINST COPERNICANISM IN THE *TRACTATUS*

Turning finally to the *Tractatus* iself, what we find is the strongest and most rigid theological attack against Copernicanism that appeared in Galileo's day. The overall impression throughout the book is that it is a blizzard of quotations from scripture and the church fathers, as Inchofer employs the standard theological method of his time in arguing by the citation of authorities. There is no need for us to directly examine this great amount of detail in the text since it is translated in appendix 1 for the reader's inspection. Out of this blizzard of texts comes a long list of *de fide* truths, which include the following:

(1) God created the firmament in the middle of the waters.
(2) The heavens are up, and the earth is down.
(3) God suspended the earth above the void.
(4) During the suffering of Christ the sun experienced a miraculous disappearance for three hours.
(5) The sun moves in a circle around the earth.
(6) The heavens have a circular motion and a spherical shape.
(7) The earth is at rest at the center of the universe.

Of course number 7 is the key claim that Inchofer needs to establish. And so it is quite significant to note that the first five claims are declared to be *de fide,* but the last two propositions on the above list are said to be only "rather probably *de fide,*"[8] which is a dramatic claim by Inchofer, and which weakens his argument considerably, as we shall see. The reason for this is that these two claims (as well as others) are not found explicitly, but only implicitly in the Scriptures, because they are partially inferred.

Like Bellarmine, Inchofer adopted a factually literal, geocentric, biblical cosmology, and thus his position is neither Aristotelian nor Copernican. In

several passages he also adopted Bellarmine's view that the heavens are neither solid nor void, but fluid in character. Thus it turns out that Inchofer not
only agreed with Bellarmine's fluid-filled universe but also with his biblical
cosmology as the true picture of the world. In addition he also forcefully extended to its fullest range Bellarmine's central exegetical notion against Foscarini, namely, that all factually literal truths in the Bible are *de fide* truths. As
a result Inchofer was able to produce a long list of *de fide* truths that not only
massively refute Copernicanism, but that also are understood as obligatory beliefs for all Catholics, including Galileo. These two conclusions are Inchofer's
main objectives in the *Tractatus*.

Inchofer's central argument then ends in chapter 12 with the following
practical advice. A good Christian cannot argue in favor of Copernicanism
since it is contrary to the faith, but the church does allow the astronomer to
assume it hypothetically to make calculations, another point taken from Bellarmine's Letter to Foscarini. To this, however, is immediately added the following warning of a danger to the faith, which gives Bellarmine's original point a
new twist.

> Furthermore in using Copernicus's calculations, one can proceed in two
> ways. The first is to use them as purely mathematical hypotheses, on which
> known true physical principles are not thought to depend in any way. The
> second is to use them as hypotheses which are taken to be the same in
> kind as true natural principles, which are known to be or are reputed to
> be such, and from which certain demonstrated conclusions are deduced.
> (Inchofer 1633; see below, p. 157)

In effect Copernicanism can be entertained only as a fictional or counterfactual
assumption, which can "save the appearances." But it cannot be taken in a realistic sense as a set of potentially true natural principles. Rather at this level
"one is allowed to discuss its principles, but only to show that they are false."
To do otherwise is to create a danger to the faith.

This in effect recognizes religious faith as the corrector of natural knowledge, a point which Luke Wadding praises in his prefatory censure of Inchofer's
book. Astronomy is thus not an autonomous discipline, but is subordinate to
revelation and theology, and must correct its conclusions and principles accordingly if a conflict arises between cosmology and the Bible. Inchofer makes this
as explicit and as authoritative as he can.

Also, in session eight of the [Fifth] Lateran Council under Leo X, in a similar vein it is ordered that, when the principles or conclusions of the philosophers deviate from the true faith, then the teachers of philosophy are obligated to correct them with the manifest truth of the Christian faith and to refute the argument of the philosophers. (Inchofer 1633; see below, p. 166)

Inchofer must have seen himself as carrying out this role when he wrote the *Tractatus*.

INCHOFER'S QUANDARY

What influenced Inchofer to modify and expand the long-standing theological category of "matters of faith and morals"? The explanation is to be found in the documents of 1616 that give an account of the operations of the Congregation of the Holy Office. In February of that year the question had arisen of whether or not Copernicanism is unorthodox. As a first step of response the cardinals of the Holy Office asked a group of their advisers, called the Consultors, for their theological opinion, or censure, of two statements. Four days later the Consultors reached the following conclusions.

(1) The sun is the center of the world, and is completely immobile by local motion.
Censure: All agreed that this proposition is foolish and absurd in philosophy and is formally heretical, because it contradicts sentences found in many places in Sacred Scripture according to the proper meaning of the words, and according to the common interpretation and understanding of the Holy Fathers and of learned theologians.

(2) The earth is not the center of the world and is not immobile, but moves as a whole and also with a diurnal motion.
Censure: All agreed that this proposition receives the same censure in philosophy; and in respect to theological truth, it is at least erroneous in faith. (Galileo 1890–1909, 19:320–21; Blackwell 1991, 120)

Claims (1) and (2) are worded somewhat clumsily, but that is partially because of the need to state the central ideas of Copernicanism in the form of specific declarative sentences that could be judged to be true or false by direct comparison with the relevant passages in scripture. The unanimous recommendations of the Consultors is that both claims are "foolish and absurd in philosophy." This means that each claim is judged to be false when judged by Aristotelian-Thomistic natural philosophy, which dominated Catholic culture at the time and which was unquestionably geocentric in its cosmology. Although this perspective provided a clear rejection, it was basically irrelevant to the case at hand, which sought a theological assessment, although the philosophical condemnation set the needed tone.

On the essential theological level, claim (1) is judged to be formally heretical because it is directly contrary to many passages in scripture whose factual literal meaning had already been clearly established in the writings of the fathers of the church and by subsequent learned exegetes. A large portion of Inchofer's *Tractatus* is devoted to settling that point in detail. So he has no special difficulty in agreeing that claim (1) is heretical. A motionless sun in the center of the world is known to be not only false but also heretical, because it is contrary to the explicit words of the Holy Spirit.

But claim (2) is more complicated. The Consultors did not say that it is formally heretical, but gave only a weaker judgment: it is "at least erroneous in faith." The reason for this is that claim (2) is not directly but only inferentially contrary to what is said in the Scriptures. In other words, if we take the explicit statements in Scripture regarding the structure of the heavens and the solar system, we cannot be absolutely certain that the conclusions that we logically deduce from them are contrary to claim (2). The reason for this is that a human error of some sort may be concealed in the inference. In a given case perhaps the inference is logically invalid, perhaps a nonscriptural premise used is false, or perhaps the argument presupposes a false suppressed premise of which the investigator is unaware. Inchofer is quite clear on this.[9] Claim (2) is not as clearly a theological heresy as is claim (1). Hence when the Holy Office, relying on the advice of the Consultors regarding claim (2), issued its Decree of 1616 against Copernicanism, it did not say that heliocentrism is a heresy, but only that it is "false and completely contrary to the divine Scriptures."

To put this in another way, the controversy in Galileo's day was not simply between two diametrically opposed views, of which one must be true and the other must be false. There were other alternatives. It could be that the sun revolves around a stationary earth at the center of the world (geocentrism); it could be that the earth revolves around the sun at the center (heliocentrism); it could be that all the planets revolve around the sun, which in turn revolves around the earth at the center (Tycho's theory); it could be that both the sun and the earth revolve around a third center (occupied by the Hearth of the Heavens, which is the Pythagorean view, often erroneously stated in the seventeenth century to be Copernican); or it could be that the biblical model, which is none of the above, is true.

Furthermore what is contrary, or contradictory, to what here?[10] Claims (1) and (2) are both compound sentences containing several statements; claim (2) makes four different statements, two affirmative and two negative. Sorting out the types of contrariety in these possible cosmologies is not an easy task, and it did not seem to occur to anyone in 1616 or in 1633 to try to do that. So the Holy Office in its *public* Decree simply stayed rather close to the general wording of its Consultors, strengthening "erroneous in faith" to "completely contrary to the Scripture" without further specifications. Thus the question of heresy was left partially hanging as to what the precise meaning of the Decree of 1616 was. Would the theological and logical ambiguity of claim (2), as to its being heretical or just simply false, be enough to justify later legal action against Galileo?

This brings us to Inchofer's primary purpose in the *Tractatus*. In the early weeks of Galileo's trial, three theologians, including Inchofer, were asked to evaluate Galileo's *Dialogue* to determine whether in the book he "teaches, defends, and holds" the Copernican doctrine. They all said "yes," but Inchofer's view was much more detailed and hostile than the other two reports, which in turn became a major reason for the final judgment of "vehement suspicion" of heresy against Galileo. By the summer of 1633, when Inchofer was finishing the writing of his *Tractatus*, the question had changed. He was then considering, apparently under official urging, why Copernicanism itself was a heresy, which was a much more basic issue.

His options were limited. By then the trial had already ended and the verdict had been published widely in Europe. Under the circumstances Inchofer could hardly question the heretical status of Copernicanism. But the Decree of 1616 had not used the word "heresy," but only the phrase "false and completely contrary to the Scripture," following the hesitations of the Holy Office's own Consultors at the time. So Inchofer faced a quandary. To provide a theological justification of Galileo's conviction, he had to provide an explanation of why Coperncanism was not merely false but also heretical, that is, contrary to a matter of faith, when even the Holy Office had hesitated in 1616 over claim (2). If he had argued, which he reasonably could have, that Copernicanism was inadequately defined by claims (1) and (2) in determining its orthodoxy, he would have ended his career. He knew that he could not fault the Holy Office for making an imprudent judgment in 1616, which would have reopened the whole question of the orthodoxy of Copernicanism immediately after the end of the trial, an unwelcome development to say the least.

Another option would have been to adopt Bellarmine's view in his Letter to Foscarini (12 April 1615) that, once the correct literal meaning of Scripture had been determined, then everything so understood becomes a matter of faith because of the fact that the Holy Spirit has said so, whether or not the subject matter of the text pertains to religious salvation, in other words, to matters of faith and morals in the traditional strict sense. Although he had a great admiration of Cardinal Bellarmine's views, Inchofer seems to have chosen not to follow him on that path, perhaps because by 1633 it had become clear that such a universal inclusion of scripture made innumerable trivial facts required matters of religious belief. As we have previously mentioned, Galileo had already made that observation in 1615 in his private notes (Blackwell 1991, 107–9).

So Inchofer's quandary had to be dealt with by a new line of defense. He decided that the notion of matters of faith and morals had to be expanded to include both explicitly revealed truths in the Scriptures and others that are inferred from the former in various ways. The notion of inferential religious truth already had, of course, a long history. But Inchofer saw that these are human inferences and as such harbor the possibility of unknown errors creeping into the process because of logical slips or unrecognized presumed premises. So he added the notion that such derivative matters of faith could be said to be highly probably true. Thus not all matters of faith are absolutely certain; some are just probable in varying degrees. This can easily be seen in his laborious classifications of various cosmological ideas as "rather probably" matters of faith. Thus in his culminating argument we find a complex string of proofs in the last paragraph of chapter 10 of the *Tractatus* to finally show that Copernicanism is a heresy because it is contrary to the probable matter of faith stated in claim (2), that is, that the earth is at rest in the center of the universe.

But would the probability status assigned by Inchofer to claim (2) be adequate to settle the issue of the theological justification of the condemnation of Galileo? The cynical reader might conclude from this that Galileo was only a probable heretic. Yet even after Galileo had died, Urban VIII's animosity toward his former friend did not lessen, as he strongly blamed him for causing a "universal scandal against Christianity by means of a damned doctrine" (Galileo 1890–1909, 18:378–79). Was Copernicanism really that much of a scandal?

Or was it that the prosecution of the case against Galileo had raised unforeseen fundamental questions about the scope of "matters of faith and morals," which Urban VIII had seen as still lingering threats to the future of the Church? For before the year 1633 had ended, the Church was already preparing defenses of the verdict.

In short Inchofer's attempt to justify the verdict of Galileo's trial led him into a hidden theological and logical ambiguity in the 1616 Decree of condemnation of Copernicanism. Since the heresy verdict was already a "fait accompli" by June 1633, Inchofer's response to this quandary was to employ the notion of "rather probable *de fide* truths."

In summary Inchofer's quandary arose as follows. According to Bellarmine anything that is *explicitly* stated in the Bible, assuming that it is understood in its correct literal sense, is not just true but is *de fide* true, because of the complete veracity of the Holy Spirit who is the speaker of the words in the scripture. This would make the denial of even very trivial statements of information in the Bible to be heretical, assuming that the literal meaning has been correctly understood. Galileo had already objected in 1615 to such a broad range of explicitly stated *de fide* truths. But Inchofer saw a different problem, namely, that the second component of Copernicanism (i.e., the earth is at rest at the center of the world) is not contrary to what is *explicitly* stated in the Bible, but only to what was *inferred* from what is explicitly stated, and error may lurk somewhere in these inferences. Inchofer had to create a special category for claims that are inferred from the wording of the Bible, and so he introduced the novel notion of "probable *de fide* truths." So the question faced by Inchofer was whether something that is contrary to what is inferred from the scriptures (i.e., the earth is at rest at the center of the world) is itself a heresy. If this cannot be established, then it is not definite that Copernicanism, even if it be false, is a heresy.

Thus it turns out that, in this specific case at least, theological thinking in 1633 regarding the central charges against Galileo had indeed developed significantly from the religious justification behind the Decree of 1616. How this was received in the Vatican we do not know. But we do know that Inchofer's justification of the verdict essentially depended on a genuine theological novelty: "probable *de fide* truths." No matter how highly probable such a truth may be, it still is open to the possibility of error. Is the notion of a "fallible matter of faith and morals" an oxymoron? If not, does it have any fruitful consequences for our understanding of the nature of religious belief? We will return to these issues in chapter 5.

Christopher Scheiner's Dilemma

Between Galileo and the Church

IN EARLY JANUARY 1634 AN ELDERLY JESUIT NAMED CHRISTOPHER
Scheiner (1573–1650) was traveling through northern Italy just after a critical
culminating point in his life. After a sojourn of ten years in Rome he was en
route to Vienna, and then two years later to the Jesuit college at Neisse[1] in Bohemia, where he had previously served as rector, and where he would remain
the rest of his life.

During that journey he must have been in a reflective mood over the events
of his career as a scientist, and especially over the events of the last few years
when he had had an influential, albeit indirect, role to play in the condemnation of Galileo by the Catholic Church. Scheiner was the premier Jesuit scientist of his era. His observational and theoretical work and voluminous writings of the previous twenty-five years had focused on mathematics, optics, and
astronomy, especially solar astronomy. He thus was particularly well qualified
to attest to the significance of the latest developments in these areas, including
Galileo's contributions. And in at least the early years of their interaction, Galileo thought highly enough of Scheiner's work on the newly discovered sunspots

to write and publish detailed assessments of the latter's observations and interpretations, with which he basically disagreed. But Galileo saw him then as a serious fellow scientist whose refutation could serve as a vehicle to present his own views.

Further, as we shall see, there is considerable evidence that indicates Scheiner was one of a small group of Jesuit scientists who not only welcomed and participated in the newly emerging mathematical sciences but also attempted to engineer a program for their acceptance by the Jesuits as an alternative to the older Aristotelian natural philosophy, which dominated in the schools of the day. If this attempt had been successful, Catholic culture may well have evolved into a form more congenial to the new scientific mentality. And as a result the subsequent unfortunate history of the relations between modern science and religion would have been very different, and certainly much less hostile. But that was not to be.

Why did this early attempt at a rapprochment between modern science and religion fail? It was due in part, of course, to the later mutual dislike, indeed sharp animosity, between Galileo and Scheiner, which increased over the years. But irascible personality characteristics alone are not enough to explain what happened. And indeed there were other less well-known but more significant factors involved, which I will delineate. These cultural and historical forces came to be focused in a special yet enigmatic way in the scientific and religious career of Christopher Scheiner. On his wintry journey from Rome back to Bohemia he may well have reflected at length, and in some bewilderment, on what these factors were and how they had converged in his life.

SCHEINER'S DISCOVERY OF SUNSPOTS

In the early years of their relationship, Galileo and Scheiner became involved in a futile priority dispute over the discovery of sunspots. Actually neither of them was "first." Because some sunspots are large enough to be visible to the naked eye, they were occasionally observed over the centuries in Europe, in the Near East, and in China. In 1607 Kepler observed a sunspot that he took to be Mercury passing in front of the sun. There were no means available to study such spots systematically, however, until after the invention of the telescope.

Then in about a two-year period during 1610–11 sunspots were "rediscovered" and studied independently by Johann Fabricius (the first to publish on them in 1611), by Thomas Harriot, by Galileo, by Scheiner, and by Domenico

Cresti da Passignani. Galileo's first observations (perhaps as early as July–August 1610, certainly by April 1611) seem to have preceded Scheiner's (April and November 1611), although the latter published his observations and interpretations first (*Three Letters on Sunspots*) in January 1612 under the pseudonym of "Apelles," while Galileo's *Letters on Sunspots* was not published until March 1613. Meanwhile it is not at all clear who may have heard reports, and when, of the work of others in this original group of investigators.[2]

Although these priority claims have since received a great deal of attention, they are much less important for our concerns than the questions relating to how sunspots were interpreted and what these interpretations reveal about the cultural forces acting at that time. To focus on these issues, let us look rather closely at Scheiner's views on these conceptual disputes.

In the spring 1611 two independent events occurred which were to determine the character of all of Scheiner's subsequent career. First, in April he happened to observe, totally by accident, that dark spots are apparently present on the surface of the sun. Second, in May the presiding General of the Society of Jesus issued a letter instructing all Jesuits that in their teaching and writing they must follow the theology of Thomas Aquinas and the philosophy of Aristotle (see appendix 2.B).

Depending on how one interprets the status of sunspots, these two developments are inconsistent with each other. For since sunspots are observed to change with some rapidity, then if they are located on the surface of the sun as they appear to be, then it follows that the sun, and by implication other celestial bodies as well, are not inalterable, as Aristotle's cosmology unequivocally and systematically claimed. In effect Scheiner happened to have in hand a critical piece of evidence that could prove (and later actually did prove) that Aristotle's cosmology was wrong in one of its fundamental claims.

We have no way of knowing whether Scheiner was immediately aware of the dilemma he faced, since he did not actually look into the phenomenon of sunspots in any detail until about half a year later. But by then, or soon afterwards, he must have seen the problem, and his writings on sunspots from that point on show that he tried to avoid a conflict between his work as a scientist and his obligations of religious obedience as a Jesuit priest. Thus in the career of this individual person we have the science-religion conflict lived out in the concrete in a particularly sharp way, and at the critical time when these issues were just beginning to be defined for centuries to come.

At the time of his discovery Scheiner had just recently arrived at the University of Ingolstadt, where he taught mathematics and Hebrew. He of course had previously completed extensive studies in the traditional philosophy and theology of his day as preparation for his ordination as a priest, along with his more specialized studies in mathematics. He tells us that in the spring of 1611 he and a friend (Johann Baptist Cysat, S.J.) were engaged in making a series of telescopic observations for the purpose of measuring and comparing the apparent diameters of the sun and the moon. In the process they happened to observe some very dark spots on the sun, but they did not look into them any further then, as they were engaged in another project.

It is important to note that this research project was not an abstract philosophical debate so typical of the day, especially within the Jesuit schools. Rather it was what we would now call a properly scientific undertaking: an attempt to make precise measurements of physical objects. Why he was interested in this particular project, and how it might have fit for him into any larger research plans, we do not know. But we do know that at that time such scientific work was actively conducted by a number of Jesuits throughout Europe. These men were the intellectual descendants of Christopher Clavius, S.J. (1537–1612), who a generation earlier had started a scientific subculture within the Jesuits.[3] So from the very beginning Scheiner was already traveling down a new and different intellectual road compared to most of his Jesuit confreres.

Scheiner also tells us that in the next October he began to make further observations of sunspots, which he described as "almost beyond belief." He probably meant this quite literally, for as he well knew, such spots on the sun simply should not be there if Aristotle's cosmology is correct. The reason for this is that Aristotle had argued with great detail and insistence that objects in the celestial realm (from the sphere of the moon on out to the end of the universe in the fixed stars) can change *only* by way of local motions, that is, by the changes of place that are clearly observed in the daily rotations in the heavens. Aristotle had insisted that his three other categories of change—quality, quantity, and substantial nature—are limited to the terrestrial realm. In the technical language of the Aristotelians, the heavens are incorruptible or inalterable, except for local motion.

Given the Aristotelian intellectual climate of the times, the question thus arose of how these sunspots should be interpreted ontologically, in other words, what status, if any, should be assigned to them in the real world. Scheiner continued his observation and study of the spots for the next two months, and he hurriedly summarized his results in a set of three letters to Mark Welser, a prominent magistrate in the city of Augsburg and a close friend of the Jesuits.

Welser then published this material in January 1612 under the title *Tres epistolae de maculis solaribus*. These three letters are of particular interest because they capture Scheiner's original sense of intellectual excitement over his important discovery, his maneuvers to interpret the new evidence in hand, and his scientific vs. abstract philosophical approach to the issue.

In these letters Scheiner has no difficulty in rejecting the possibility that the sunspots are deceptions due either to the human eye or to the telescope. Also they are not located nearby in the earth's atmosphere, since they exhibit no parallax. Hence they must be located either on the sun or in the near vicinity of the sun. At this point he makes the critical move to reject the former view primarily for the following reason. If the spots were on the sun, then the sun must rotate on its axis, since all the spots drift slowly in one direction across the face of the sun. As a result when they disappear around one side of the sun, they should reappear about two weeks later rising on the other side in the same position relative to each other and to the sun. But this is not what he observed. Rather they simply disappear, and different new spots are found instead. Scheiner does not seem to have considered the notion that the spots themselves are impermanent, which would make the sun even more qualitatively variable.

At any rate Scheiner finally concluded that the spots are caused by the light of the sun being blocked by celestial bodies (which he calls "stars") or by aggregates of such bodies. These stars move across the sun near its surface, and thereby cause "mini-eclipses" of the sun's light. He held firm to this basic view throughout his original debate with Galileo on the matter. In his later *Letters on Sunspots,* Galileo, in turn, interpreted the spots as either located on the surface of the rotating sun or else very close to it, like huge variable clouds rotating with the sun. His best argument is based on careful observations that the foreshortening effect visible in the varying shape of a sunspot as it moves across the sun is the same as what would be seen in an object attached to a rotating sphere.[4] There are, of course, numerous other differences and arguments in the views defended by Scheiner and Galileo, but the main points for our concerns are the ones mentioned above. The key consequence of Scheiner's interpretation is that the heavens near the sun are now understood to be populated with a very large number of previously unseen new stars, but all of them, as well as the sun itself, preserve their traditional Aristotelian status of being inalterable, except for local motion.

Scheiner's view was a reasonable one. Also his personal open-mindedness and lack of ill-feeling toward Galileo at this point can be seen in his response one year later after Galileo's criticisms became known to him. In a later book reviewing recent developments in astronomy, Scheiner softened his position as he summarized his view, on what had by then become the sunspot controversy, in words that were hardly those of a zealot.

> They [sunspots] are black bodies which have various and erratic motions across the sun; neither their nature nor their number has yet been determined. They are so close to the sun that they cannot be separated from it by the senses. . . . Whether they are stars is still a matter of dispute; time will tell. Examine the drawings of Apelles, study Galileo's *Letters on Sunspots*. Many things are to be expected in time. (Locher 1614, 66)[5]

Granting all this, and knowing now with the easy wisdom of hindsight that Scheiner was wrong in his interpretation of the sunspots, then why did he miss the mark? The traditional explanations are that his scientific work itself was defective (e.g., his drawings of the spots were inaccurate; he failed to appreciate the impermanence of the spots) or he was so subliminally influenced by the accepted Aristotelianism of the day that he read that into his evidence, a very understandable type of thinking that we all experience.

But the situation was considerably more complicated because of another factor which contemporary scholars have generally not taken into account. Scheiner was also under a very specific religious pressure, which arose from his role as a Jesuit, and which he must have had consciously in mind.

THE GENERAL ORDERS AND THE SOLDIERS TAKE COVER

This takes us back to the second major event that affected Scheiner's life in the spring of 1611. In May the presiding General of the Society of Jesus, Claudio Aquaviva (1543–1615), issued a letter of instruction (see appendix 2.B) for all Jesuits to observe. The occasion of the letter was a strong concern in the General's mind that the Society was faced with a problem of losing its unity and effectiveness, its commitment to its original goals of serving the Church by protecting the faith, and possibly the good will of Pope Paul V.

There were two reasons for this situation. First, the number of Jesuits and of the schools they sponsored had increased very considerably during the previous quarter century. As a result Jesuit intellectual activities and published treatises had also expanded enormously, both in numbers and in subject matter. This in turn entailed a great deal of academic pluralism and internal differences of opinion.

Second, Aquaviva was concerned with arousing the Pope's disfavor because Jesuit writings continued to appear on the topic of how to reconcile human freedom with God's infinite causal power and foreknowledge of human actions. This thorny theological problem had been the subject of a long and bitter debate between the Jesuits and the Dominicans that began in 1588. The rivalries were so strong and the debates so inconclusive that in 1607 Pope Paul V, in an effort to preserve unity within the Church, declared in effect a moratorium on the issue, which he reserved for his own later decision (neither he nor any subsequent Pope ever actually attempted to resolve the matter).[6] Under these circumstances Aquaviva was understandably convinced that any additional innovative Jesuit treatises on the topic were not welcomed. The General's reasons for issuing the letter seem to have been of a purely religious and theological nature, and not at all focused on the recently emerging Jesuit scientific work as such, although that was also affected, as we shall see.

To remedy the situation Aquaviva prescribed a heavy dose of what he called "solid and uniform teaching." This medicine was nothing new. It was rather a reaffirmation of the original intellectual orientation of the Jesuits, as stated in the *Constitutions* of the Society of Jesus. The General also depended heavily on a resolution (see appendix 2.A) that had been adopted at a general meeting of the Jesuits in 1594. This document specified that Jesuits are required to teach the doctrines of Thomas Aquinas in theology and of Aristotle in philosophy. Regarding the latter the only exceptions stated arise when Aristotle's views are contrary to what is commonly taught in all the schools and when his views are contrary to orthodox religious faith. In the latter case the Jesuit had a special obligation to refute Aristotle's arguments. Needless to say, the conservative "taught in all the schools" requirement was an attempt to discourage intellectual innovation, which ironically had in many ways contributed significantly to the recognized success of the Jesuit schools.

Aquaviva's letter explicitly reaffirmed the above doctrinal requirements and went on to add more specifics. The standard of "solid and uniform teaching" required more than simply the obligation to teach what is true. In addition to freedom from error, one's conclusions, principles, and methods of proof should conform to those found in Aquinas and Aristotle. Moreover Jesuit "censors" were instructed to be more critical and vigilant in applying the standards of "solid and uniform teaching." These censors constituted a system of internal prepublication review of Jesuit treatises (similar to, but stronger in authority than, present-day peer-review procedures) which had been instituted years earlier to avoid embarrassments for the Society. Thus the General signaled that his orders were to be taken seriously.

Aquaviva's 1611 letter also ordered a review and evaluation process. Each regional group of Jesuits was asked to submit assessments of, and further recommendations for, the announced policy. As a result, two and a half years later, in December 1613, Aquaviva issued a follow-up letter (see appendix 2.C) which seems to be even more stern than the first one, especially on its rejection of innovations in philosophy and theology. The Jesuit scholar is not to introduce new ideas on his own initiative. What then was Scheiner to do, if anything, about his newly discovered sunspots, which seemed to indicate that Aristotle was wrong on his claim that the heavens are inalterable?

The disciplinary penalties are made specific. Those Jesuits who disagree with the approved teachings of Aquinas in theology and of Aristotle in philosophy should not be appointed to teaching chairs in Jesuit schools and should be removed from teaching if already appointed. Those teaching contrary views should retract them immediately and not even wait until the end of the course. Even more strongly Aquaviva appeals to the particularly strict Jesuit vow of obedience[7] ("we have schools, as is pleasing to God, in which anything which falls outside of the laws of obedience is simply unacceptable"). The academic status and future scholarly writing of the Jesuit frontline troops were directly at stake. Again, what was Scheiner to do?

As a result of all this a very peculiar situation was created for the Jesuit scientists of the day. For them the result of Aquaviva's letters was a case of "unintended and unforeseen consequences." However harsh or unwise the General's orders may have been, they were clearly intended to deal with the Society's internal problems of unity of purpose and of viewpoint as a religious institution, and to avoid the displeasure of the Pope over a thorny theological dispute. The latter is particularly clear from the Decree on Efficacious Grace (see appendix 2.C) appended to Aquaviva's second letter. But the stated remedy to deal with these matters was a strict reaffirmation of Thomist theology and Aristotelian philosophy without room to maneuver, and this happened precisely at the time when a subgroup of Jesuit scientists was beginning to move beyond Aristotle. The General was certainly not primarily concerned, perhaps not even significantly concerned, at that time about the status of philosophy among the

Jesuits. The Copernican challenge to geocentric cosmology was not yet a major issue in the Catholic world,[8] although it was waiting only a few years away in 1616. But another intellectual crisis for Jesuit scientists had already been generated *before* that date, and it arose for nonscientific reasons. How could they reconcile their dual roles of being both priest and scientist?

What were the General's soldiers to do under these circumstances? Open opposition would have been out of the question. Personal concealment and avoiding or postponing the issue were the preferred strategies. There is more than enough surviving evidence to verify this. For example, consider the case of Christopher Grienberger, S.J. (1561–1636), who succeeded Clavius in the chair of mathematics at the Collegio Romano in 1612. In November of that year he is reported to have said, in comparing the views of Galileo and Scheiner on sunspots:

> Even though he [Grienberger] knows that Apelles is a Jesuit, he still agrees much more with you [Galileo] than with Apelles. . . . However, as a child of obedience, he does not dare to say what he thinks. (Faber's Letter to Galileo, 23 November 1612; Galileo 1890–1909, 11:434)

In June 1614, six months after Aquaviva's second letter, Grienberger was present at a lecture and experimental demonstration at the Collegio Romano in which Giovanni Bardi explained Galileo's views on how bodies float on water. Bardi then wrote to Galileo,

> Fr. Grienberger told me that if the topic had not been treated by Aristotle (with whom, by order of the General, the Jesuits cannot disagree in any way but rather are obliged always to defend), he would have spoken more positively about the experiments because he was very favorably impressed by them. He also told me that he was not surprised that I disagreed with Aristotle, because Aristotle is clearly also wrong in regard to what you [i.e., Galileo] once told me about two weights falling faster or slower. (Bardi's Letter to Galileo, 20 June 1614; Galileo 1890–1909, 12:76)

Another Jesuit, Giuseppe Biancani (1566–1624), also discussed this same scientific issue in a book submitted to the Jesuit censors in February 1615. One of the censors made the following recommendation about a proposed emendation.

> The addition to Biancani's book about bodies moving in water should not be published since it is an attack on Aristotle. . . . It does not seem to be either proper or useful for the books of our members to contain the ideas of Galileo, especially when they are contrary to Aristotle. (Blackwell 1991, 150)

Exactly one year later Biancani presented another book, his *Sphaera mundi* (1620), to the Jesuit censors. One of them happened to be Grienberger, who laments that Jesuit mathematicians are not free to take part in a much needed rethinking of the overall theory of the heavens.

> It seems to me that the time has now come for a greater degree of freedom of thought to be given to both mathematicians and philosophers on this matter, for the liquidity and corruptibility of the heavens are not absolutely contrary to theology or to philosophy and even much less to mathematics. . . . Up to now his [Biancani's] hands have been tied, as have ours. Thus he has digressed into many less important topics when he was not allowed to think freely about what is required. (Blackwell 1991, 152)

Grienberger's hands seem to have remained tied, since he published only on purely mathematical topics after Aquaviva's orders were issued.

Another telling remark is found in a letter written less than three months after the end of Galileo's trial. Again the impact of the General's insistence on "solid and uniform teaching" is evident.

> Fr. Malapert and Fr. Clavius himself did not disapprove of the opinion of Copernicus, and were not very far from it, but they had been pressured and obliged to write in favor of the common views of Aristotle, which even Fr. Scheiner himself supported only because of force and obedience. (Niccolò Fabri di Peiresc's Letter to Pietro Gassendi, 6–10 September 1633; Galileo 1890–1909, 15:254)

In the light of such evidence there seems to be little room for doubt that Aquaviva's two letters mandating the teachings of Aquinas in theology and especially of Aristotle in philosophy constituted a major obstruction for the work of Jesuit scientists, who ironically were just beginning to move past Aristotle. Scheiner, of course, was no exception. The last quotation above even makes us wonder whether his stated views were really his own, at least in the later years.

We might note in passing that in 1612 Scheiner used another strategy to avoid confrontation with the General's orders. As mentioned earlier he published his views on sunspots under a pseudonym. Scheiner himself tells us years later that his religious superiors were concerned about the embarrassment to the Jesuits that might result if his new findings, so contrary to the received views in philosophy, were published too hurriedly and would later need to be retracted. Their advice was as follows:

> My superiors were of the opinion that I should proceed carefully and slowly, until these phenomena were corroborated by others who had come upon the same evidence; that I should not depart from the common path of the philosophers without evidence to the contrary; and that my observations in the letters sent to Welser should not be published under my own name. In this way a greater freedom of thought would be available to all in the future, and no one would have ill feelings. As a result of these cautions, the letters which were published were many fewer than I would otherwise have written to Welser; they were published under the name of Apelles; and many things proper to the philosophers are discussed in them besides my own views. (Scheiner 1626–30, part I, chap. 2)

In this later public statement of what happened no mention is made of the requirement of "solid and uniform teaching." But it is virtually certain that it must have been taken into account internally as Scheiner's Jesuit superiors made these decisions just six months after Aquaviva's first letter. Clearly a way was sought to claim credit for the Jesuits if Scheiner's discoveries were correct, but to avoid criticism if they were not. As it turned out, within one year it became known that Apelles was a Jesuit, and within two years that he was Scheiner. At any rate the General's orders had an almost immediate effect of discouraging innovation among Jesuit scientists.

Scheiner actually used two pseudonyms in his 1612 writings on sunspots: *Apelles latens post tabulam* ("Apelles hiding behind his painting") and *Ulysses sub Aiacis clypeo* ("Ulysses behind the shield of Ajax").[9] In each case Scheiner identified himself as a famous person under a disguise. Apelles was the foremost painter in ancient Greece, who is reported to have concealed himself behind his paintings in order to listen to, and benefit from, the criticisms of expert (but not of unknowledgeable) viewers.[10] And indeed Scheiner did solicit and take account of Galileo's criticisms of his work on sunspots. Ulysses, the man of reason and the heroic survivor of the Trojan War, appears in the guise of Ajax,[11] who was more a man of action and a furious warrior. Are we thus to think that Scheiner thought of himself personally as being reasonable rather than pugnacious in his early debates with Galileo on sunspots? At any rate this soldier literally took cover after the General's orders were issued.

At the end of the 1611–13 exchanges between Galileo and Scheiner on sunspots, the debate had been kept well within the bounds for that time of friendly disagreement. Galileo had become a bit more testy toward the end, and Scheiner became somewhat less sure of his views. The bitterness of their later exchanges had not yet appeared. Meanwhile the friends of each of them counseled a more aggressive stance, especially in regard to claiming priority for the discovery of sunspots.

THE MIX-UP OVER THE "MIXED SCIENCES"

Up to this point I have on several occasions mentioned that when Scheiner discovered sunspots, some Jesuit scientists were on the verge of moving past the Aristotelian conception of science. Recent research on this topic has shown that this movement began in the 1580s under the leadership of Christopher Clavius and his Jesuit students at the Collegio Romano. I will not attempt to reconstruct here the history of this development, which has already been done very well by others,[12] but will focus rather on the conceptual issues involved. This is the story of Aristotle's original notion of what a demonstrative science is and its evolution in the hands of the later medieval philosophers.

In his *Prior Analytics,* Aristotle succeeded in identifying and accounting for all of the valid and invalid forms of the categorical syllogism, a major achievement in its own right. From this investigation it became apparent that the truth or falsity of the statements in a proof has nothing to do with its validity or invalidity. This may well have prompted Aristotle to carry the discussion on to the next step and to ask what further conditions must be satisfied for a valid syllogism to produce a conclusion that was not merely true as a matter of fact, but of which it is proven that it *must* be true. At any rate this was the central goal to be attained by Aristotelian "demonstrative science."

These conditions are presented in great and demanding detail in his *Posterior Analytics* (bk. 1, chaps. 1–10). They include the requirements that the premises be universal statements, that they be necessary truths about the actually existing world, that they state the actual causes of the objects and events in that world, that the middle term of the syllogism expresses the essential definition of the subject matter of the science involved, and that these premisses be "first" in the sense that they cannot themselves be proven, or deduced, from other more fundamental and more universal premisses. The net result of all this was a model of demonstrative science that consists of a set of universal statements arranged in a continuous deductive chain hanging down from a small set of indemonstrable first premisses.

The question that thus naturally arises, of course, is what is the origin and the justification of the required absolutely first premisses. One could take the Platonic option here, and say that they are grounded in a priori innate concepts in some way, thus arriving at some version of an idealistic conception of science. But Aristotle explicitly rejected this. Rather, as an empiricist he argued that all human knowledge is derived from the senses. As a result the deductive phase of scientific proof must be preceded by a quasi-inductive generative phase that produces the critical first principles. He, of course, knew well that simple enumerative induction, being logically invalid, could not do the job. The *Posterior Analytics* ends instead with an enigmatic appeal to an intellectual grasp of the essential structures residing in real objects as the origin of a science's first principles. Science thus consists of two phases: one generative of the fundamental principles, the other demonstrative of the conclusions that follow from them.

The Aristotelian view of science thus presupposes that individual objects in the physical world are so constituted that they belong to natural classes, that is, species and genera; that each species is determined by a common essence or nature or substance (as distinct from the myriad of accidental properties of the individual); and that the human mind has the power to know these substantial natures by generalization and abstraction from concrete cases. This knowledge is part of the Aristotelian claim that the sciences explain the objects and events of the world in terms of their causes (the formal, material, efficient, and final causes, the first two of which define the essence.) This notion of science, which served as an epistemological ideal for many centuries, stands like a shadowy skeleton within the various surviving writings of Aristotle on what we would now call physical science. However it is clear that he was not able to meet fully his own ideal. A much better instance is found in Euclid's *Elements,* but that of course was presented as a purely mathematical, not a physical, science.

Aristotle himself thought that his model applied both to the various branches of what he called "physics" (i.e., the sciences of nature) on the one hand, and to metaphysics on the other hand, which deals in his view with the realm of the immaterial and the unchangeable. In between, as it were, there was a third group of theoretical sciences (geometry and arithmetic) that deal with intermediate objects like lines and figures, which exist in material bodies, but which are unchangeable and fixed because they are considered in abstraction from such embodiment.

Because of this latter factor these mathematical entities were not understood to be ontologically real as such, and thus they do not meet all the requirements in Aristotle's model of science. In effect the objects of the mathematical sciences are idealizations constructed by the human mind. For example, geometrical lines and circles are perfectly straight or round, although physical lines and circles are not. As a result of this peculiar status of mathematical objects, arithmetic and geometry fell short of the ontological and causal requirements of the Aristotelian model of a demonstrative science. For this reason mathematics was thereafter assigned a special secondary status as a science.

In time this conception of the sciences came to dominate Islamic and Western Christian culture. But gradually some further complications arose. In between the physical sciences and mathematics as described above, there emerged a set of other disciplines, like astronomy, optics, mechanics, and music, that deal with physical matters, but do so by borrowing their needed explanatory principles from mathematics. Because of this dual input, they were called "mixed sciences." Most importantly for our concerns, they were understood at that time to be non-ontological and noncausal in import, precisely because the explanatory first principles in these disciplines were borrowed from mathematics.

If this view of the mixed sciences be granted, then two consequences follow. First these disciplines satisfy the logical requirements of Aristotle's model of a science, since they can be organized into a series of connected deductive proofs derived from one set of first premises. Second, since the latter have no causal or ontological status in the objective real world, the assertions of a mixed science cannot be taken as having a realistic import. Rather they are understood as "hypothetical" in Bellarmine's sense of that word; that is, if the mathematical principles are assumed for the sake of argument, then the observed phenomena would be accounted for as their consequences. The appearances would thus be "saved" in the sense that they are organized systematically within the proposed deductive structure, which in turn supplies methods for calculations, but nothing more.

This meaning of "hypothetical" is not the same as the modern sense of an "hypothesis," that is, a reasonable guess, after some evidence is available, of what a general empirical law may be, and which is worthy of further research to test its objective truth value. Rather "hypothetical" regarding a mixed science meant that a given assumption is nonobjective, that is, it is *not* true of the world as such. The reason for this is that the mathematical objects in its premises were considered to be abstract idealizations, not really existent structures residing in the objective world. This nonobjective meaning of the term is what the churchmen of Galileo's day regularly had in mind when they advised the astronomers to treat Copernicanism "hypothetically."[13] It was advice to interpret the new theory as a case of a nonrealistic mixed science.

Granting this, Jesuit astronomers (who like all astronomers were usually called "mathematicians" in the sixteenth and seventeenth centuries) were thus indeed expected to be highly mathematical in their theories. That was their proper method. But their results could have no ontological import. Precisely because of their mathematical methods, they were merely able to "save the appearances" and to make calculations. Thus their traditional devices, such as epicycles, eccentrics, and equants, were not to be understood as in any way delineating the actual structure of the cosmos. That job was reserved to the natural philosophers of the day, who claimed to possess Aristotelian-type intellectual understandings of the essential structure of the heavens and its constituent bodies.

This division of the territorial boundaries between the natural philosophers and the astronomers clearly gave much superior authority to the former group within the Jesuits. Clavius and his influential students made a concerted effort from the 1580s into the 1620s to claim their independence from the natural philosophers, but in the long run they did not succeed. Their tactics consisted basically in trying to argue that, in the generative phase of the growth of science, mathematical intelligibilities were to be found in the objective world. They argued, in effect, that close observation of the physical world showed that quantitative structures and relationships (and not only qualitative essences) are directly revealed to the senses as real.

If this be granted, then the status of mathematical objects as merely abstract idealizations was to be rejected in favor of viewing them more realistically. This approach of the Jesuit mathematicians was still Aristotelian in its overview of science, but in its generative phase it replaced or supplemented essential natures with mathematical structures as the type of intelligibilities found in the real world. They were moving in the same direction as Galileo, who put the matter succinctly in his famous comment, "The book of nature is written in the language of mathematics." Seeing nature in that way was critically important for the birth of the modern sciences. Although they disagreed on the interpretation of sunspots, both Galileo and Scheiner attempted to draw realistic conclusions from their mathematical analyses of the telescopic observations. They were thus working on the same level, and both were opposed to the traditional natural philosophers of the day.

Thus it now appears to be rather clear that a group of Jesuit mathematicians were actively working to transcend Aristotelian philosophy of nature by developing a new conception of a mathematicized natural science. If they had succeeded, one wonders how this would have influenced the subsequent history of the relations between science and religion. The effects would have undoubtedly been very constructive. An innovative synthesis of the traditional religious culture with the newly emerging sciences, analogous to the merging of Aristotelianism itself with medieval theology in the thirteenth century, may have been at least a possibility.

But they failed. Why? One reason, no doubt, was the internal opposition from the Jesuit natural philosophers who would have lost their hegemony.[14] It is clear that a struggle occurred between these two groups of Jesuits over their places in the schools, and that the innovative mathematicians were always a small minority. Another clear reason was the impact of Aquaviva's forceful and unfortunate letters requiring Jesuit teachers and scholars to return to strict Aristotelianism in philosophy. The Jesuit mathematicians were struck down as innocent bystanders before their case was fully made. Even if they had been attracted to the new paths being explored by Galileo, they could not continue that journey without elaborate disguises.

Although the issue had not been decided by the year 1616, developments at that time settled the matter permanently. With the condemnation of Copernicanism as "false and completely contrary to the Holy Scripture," the Jesuit astronomers had little room to maneuver.

THE ROAD TO ROME AND TO BITTERNESS

Scheiner's exchanges with Galileo over the question of sunspots had come to an end by 1614. For the next ten years he carried on his teaching and research at various German Jesuit colleges, moving to Innsbruck in 1618, to Freiburg im Breisgau in 1621, and to Nissa in 1622. We know rather little about his specific activities during those years, although he must have laboriously continued his observations of the sun since he later published a massive study on that topic in his *Rosa ursina sive sol* (1626–30). This book established him as the leading authority of his day on solar astronomy. However he did become involved in another wider dispute between the Jesuits and Galileo in the years 1618–24. This time the quarrel turned personal, very bitter, and hostile on both sides, and ended with a permanent breaking of relations.

What happened briefly was the following.[15] In October and November 1618 three strikingly noticeable comets made their appearance. They were taken in the general population to portend great evils. And indeed the disastrous Thirty Years War between the Catholics and the Protestants began that same year, bringing massive destruction to much of central Europe.

For the astronomers, however, the issue was the nature of comets and where they are located. Early the next year the Jesuit Orazio Grassi (1590–1654), who then occupied the chair of mathematics at the Collegio Romano, published his *Disputatio* in which he proposed the anti-Aristotelian and pro-Tychonic view that the comets were impermanent bodies located far out beyond the moon. His arguments were scientifically questionable, but the tone of this treatise was not hostile and did not criticize Galileo in any way. The author of the *Disputatio* was not identified as Grassi, but only as an anonymous Jesuit, who later adopted the pseudonym of "Sarsi" as the controversy with Galileo got underway. In effect, this was another case of "hiding behind the picture."

By 1619 the atmosphere had, however, changed very considerably from what it was in 1612. At the earlier date both Galileo and Scheiner disagreed respectfully and avoided personal criticism. Now Galileo's Lincean friends advised blunt and forceful opposition; so did Grassi's Jesuit friends. And both followed this advice. We will focus shortly on the reasons for this souring of relations. As a result Galileo published his *Discourse on Comets* later that year in which he sharply and sarcastically criticized the anonymous author's arguments and the competence of Jesuit scientists in general. His *Discourse*, by the way, was published under the name of his former student Mario Guiducci, even though we now know that most of the manuscript is in Galileo's handwriting. Both sides took advantage of concealment.

This initial clash was followed by a series of increasingly bitter replies and counterreplies, and ended only with Galileo's *Assayer* (1623) and Grassi's (writing

as Sarsi) *Ratio ponderum* (1626). The former is well-known as Galileo's mani-
festo on the nature of science and its methodology, but is also a classic of biting
polemical criticism. The once friendly relationship between Galileo and the
Jesuits was shattered without hope of reconciliation.

Scheiner was pulled into this dispute because of a gratuitously offensive
insult about him in the *Discourse on Comets*. Guiducci says sarcastically

> I could not care less for the honor of having been a good plagiarist, a role
> which has been taken up by those who make themselves out to be the dis-
> coverers of his [Galileo's] other ideas, and to pretend that they are Apelles
> when, with their drab colors and poorly drawn designs, they clearly reveal
> that in painting they are not even comparable to artists of mediocre talent.
> (Galileo 1890–1909, 6:47–48)

Scheiner was furious over this remark, and is reported to have said that
he would repay Galileo "in the same coin."[16] This was the beginning of charges
of plagiarism by both sides, which darkened the later stages of the Galileo-
Scheiner dispute. Stillman Drake has suggested that Scheiner began his "re-
payment" by sending material to Grassi to use in his replies to Galileo (Drake
1978, 276), but there is no direct evidence for this. Be that as it may, Scheiner
did extensively circulate the "same coin" in two of his later books. To make matters worse, Galileo repeated the plagiarism charge in the *Assayer*.

> How many men attacked my *Letters on Sunspots,* and under what dis-
> guises! . . . [S]ome, overwhelmed and convinced by my arguments, at-
> tempted to rob me of that glory which was mine, pretending not to have
> seen my writings and trying to represent themselves as the original discov-
> erers of these impressive marvels.[17]

Because the chronology of their publications clearly gives priority to Scheiner,
Galileo may not have had him in mind here,[18] but Scheiner certainly did take
it as directed at him.

Be that as it may, if we now step back from the details, it is clear that Galileo's relationship with the Jesuits had undergone a complete reversal. The
high point of friendliness was reached in May 1611 when Galileo was feted
and honored at a special academic colloquium[19] held at the Collegio Romano
in recognition of his initial telescopic discoveries in his *Sidereus nuncius* (pub-

lished in 1610). The low point of hostility was reached twelve years later in the *Assayer* with a complete rupture of the relationship. And the low point remained permanent.

Why did that happen? What factors caused this breakdown of friendship? One possible factor sometimes mentioned is what had initially happened between Galileo and Scheiner regarding the discovery and interpretation of sunspots. But if so, that would have had a relatively minor influence, since Galileo and Scheiner disagreed then more cordially than hostilely, as we have seen. The priority of discovery dispute became prominent and nasty only years later. Even so, why would Galileo choose to alienate all the Jesuits, and not just Scheiner, over this issue? Would that have been worth the price?

A much more substantive factor was the Church's condemnation of Copernicanism on 5 March 1616 as "false and completely contrary to the Holy Scriptures." Galileo, by then a clearly identified advocate of the new astronomy, knew of course that at least some of the Jesuit astronomers, for example, Grienberger and Paul Guldin, were also quite sympathetic to Copernicanism. The venerable Fr. Christopher Clavius had said in print as early as 1611 that Galileo's observations with the telescope established a need to rethink the arrangement of the orbs in order to save the phenomena. Galileo correctly recognized that the Jesuit astronomers were the most experienced and well-informed members of the Church on these matters. He must have had strong expectations that they would have opposed the condemnation of Copernicanism both before and after the fact.

But these hopes were dashed. The Jesuit astronomers retreated. The very next day after the Congregation of the Index approved the condemnation of Copernicanism, Biancani was put on notice by Grienberger that the manuscript of his latest book, *Sphaera mundi*, would have to be changed to fit the new decree before it could be published.[20] The next year Guldin was transferred out of Italy. Also in 1617 Grienberger was replaced in the chair of mathematics by the much less sympathetic Grassi. Thereafter Jesuit books on astronomy regularly adopted the practice of reviewing all the old and new theories for the reader's education, but then of advocating Tycho's version of geocentrism as the best of the alternatives. It was now not just Scheiner, but all Jesuit astronomers, who had disappointed Galileo. His best potential scientific allies within the church had left him isolated in the debates over Copernicanism. The Jesuits in general had lost their status for him as intellectual leaders.

But there was also a third factor, not usually discussed in the literature on Galileo, which in my opinion unfortunately moved his criticisms of the Jesuits deeper to the personal and the bitter level. He lost faith in their intellectual integrity as a group. This at least would explain the depth of hostility in his otherwise gratuitous insults to Scheiner and his lampooning of Grassi in the controversy over comets.

Consider for a moment the following situation, which indeed may well have been the case. Galileo could easily have become acquainted by that time with the content of Aquaviva's two letters ordering intellectual conformity with Aristotelianism. Independence of thought and innovative ideas are banned. What respect would he have had for the Jesuit astronomers then? To make matters worse, he quite likely was aware (it was certainly public knowledge at the time) of the Jesuit ideal of triple religious obedience, in which not only one's actions, but also one's choosing and even one's thinking, should be the same as that of the religious authority.[21] Coupling that with Aquaviva's orders, what would the consequences be for the credibility of the Jesuit scientists?

Did they really think that Copernicanism is false, or were they merely saying what was expected of them? Did they really prefer Tycho's version of geocentrism? How could one tell? What did Sarsi (or Grassi) really mean in the middle of the comet controversy when he said, "I wish to say that here my whole desire is nothing less than to champion the conclusions of Aristotle"?[22] Was this actually his own personal objective, or an ironic "double entendre" reflecting the role he was playing? A severe credibility problem obviously arises under these circumstances. Do persons mean what they appear to say? If not, intellectual integrity has become an issue. At any rate if this scenario be granted, it at least explains why in 1619 Galileo's criticisms of the Jesuits became so personal and so harsh.

Consider also the personal psychological conflict this would have caused for the conscientious Jesuit astronomer. He was forced to try to adjust his intellectual convictions with his lifelong religious commitment as a Jesuit. If this is kept in mind, then a number of things said by or about various Jesuits later in the Galileo affair make more sense, as we shall see shortly. At any rate a very frustrating situation was created for them, and Christopher Scheiner was about to step into that dilemma.

In the autumn of 1624 and in the midst of the ravages of the Thirty Years War, Scheiner accompanied the Archduke of Austria, who was enroute to Spain, as far as Genoa, and then continued on to Rome.[23] He was clearly the best Jesuit astronomer of his generation. Was a bigger gun being brought in to replace Grassi, who had lost the battle of words with Galileo?

The Denouement in Rome

For the next ten years Scheiner lived and worked at the Collegio Romano. It was an exciting time to live in Rome, although we know very little about Scheiner's personal activities during that period. Pope Urban VIII, who had just been elected, was a genuine intellectual in his own right and, as expected, gave strong support to literature and the arts. His nephew, Cardinal Francesco Barberini, began a major architectural rebuilding of the city, which has given Rome much of its modern ambience.

More importantly the Galileo trial took place during Scheiner's last year in Rome, although we have no concrete evidence of whether or how he may have been personally involved. Largely because of the prominence of the animosity between Galileo and Scheiner, many writers have speculated about his role in bringing about Galileo's downfall, but it has remained speculation. Scheiner did have a scientific influence on Melchior Inchofer, the sole Jesuit who participated directly in the trial. As we have seen, Inchofer was a member of the Special Commission, which decided in the fall of 1632 that there were grounds for a trial and later advised the court on whether Galileo's *Dialogue* violated Church directives. In his personally signed report in the latter capacity Inchofer said, "Galileo's principal aim was to attack Father Christopher Scheiner, who more recently than anyone else had written against the Copernicans" (Finocchiaro 1989, 265). Viewing the *Dialogue* that way was certainly Scheiner's judgment also.

We know a great deal about Scheiner's scientific work and writing during his stay in Rome. Since he first discovered sunspots in 1611, Scheiner had carefully and regularly observed and recorded such phenomena over the years, so that by the late 1620s he was the world's best informed authority on solar astronomy. He had written up his findings in a truly immense volume entitled *Rosa ursina sive sol* (1626–30). Ill feelings between Galileo and Scheiner had not abated, as is clear from Galileo's comment when he learned of Scheiner's impending publication.

> I hear . . . that the fictitious Apelles is printing in Bracciano a long treatise on sunspots; and the fact that it is long makes me very doubtful as to whether it is not full of blunders, which, being as they were infinite, can

soil many pages, where there is little place for the truth; and I am certain that, if he will say anything other than what I have already said in my sunspot letters, he will be speaking only vanities and lies.[24]

Galileo's view that his rival could have nothing new to say was quite wrong.

Unfortunately the scientific message of Scheiner's *Rosa ursina* has been overwhelmed by the extremely bitter and sustained diatribe against Galileo in its first sixty-six pages. Nearly every detail of their exchanges in 1611–13 is rehearsed, all with the goal of establishing Scheiner's priority in the discovery of sunspots. He had said in 1619 that he would repay Galileo "in the same coin." This indeed was considerable overpayment.

Nevertheless Scheiner had made a good deal of progress scientifically. The *Rosa ursina* contains hundreds of pages of information on the construction and use of telescopes and of data on sunspot observations made over many years. Furthermore Scheiner now agreed with Galileo that the spots are located on the surface of the sun, "like ships on the sea," and that the sun rotates on its axis once every twenty-seven-plus days.[25] He was able also to make a very accurate measurement for that time of the angle at which the axis of the sun's near monthly rotation is tilted away from the perpendicular to the plane of the ecliptic.[26] But most importantly he was able to determine that the apparently curved paths of the sunspots vary over a one-year cycle, with the curvature facing in opposite directions and reaching a maximum in March and September, and a minimum in June and December.[27] Only an astronomer who had made regular sunspot observations throughout the year (i.e., Scheiner, but not Galileo) could have determined these annual variations of curvature, which, as we shall see, played a large role in his continuing dispute with Galileo. Scheiner's talents as a first-rate scientist are illustrated by the accuracy and persistence of his solar observations, especially when we recall that he had to work with the first generation of telescopes ever to be constructed, many of them by himself, and that all of his data had to be recorded by hand drawings, many of which are published in his *Rosa ursina sive sol.*

Of course Scheiner advocated an earth-centered astronomy in this book, but whether he did so out of personal conviction or out of religious obedience is not known. One reason for doubt here relates to the complexities he encountered in trying to explain the annual variations of curvature in the paths of sunspots. For if the earth is totally at rest at the center of the universe, then to explain these phenomena, multiple simultaneous motions with some highly coincidental properties must be assigned to the sun: its daily rotation around the earth, its annual rotation along the ecliptic, its near-monthly rotation on its own axis, and the oddity of exactly the earth's annual period for the poles of the sun's axis to rotate around the perpendicular to the plane of the ecliptic. He was too good of an astronomer not to have seen that these phenomena are more simply explained in the Copernican model by the daily and annual motions of the earth, plus the sun's near-monthly rotation on its own axis which is tilted to the plane of the ecliptic.

These phenomena can, of course, be explained in either way, and Galileo's later appeal to the principle of simplicity here was certainly not conclusive. But Scheiner may still have personally favored the heliocentric view of the universe for this reason. At any rate this would explain why several of his contemporaries, including some Jesuits, have said that he was secretly a Copernican.[28]

The *Rosa ursina* ends with a long and explicitly anti-Aristotelian argument to the effect that the stars and the heavens are composed of elemental fire, that they are corruptible in nature, and that they are fluid and not solid in their constitution. The incorruptibility of the heavens, so important to Scheiner in 1612, is now explicitly rejected. Less than two decades after Aquaviva's prescription of "solid and uniform teaching" for Jesuits (a rule that was to be reasserted twice in the eighteenth century), Scheiner argued vigorously to the contrary in cosmology. As usual his book had been internally reviewed by the Jesuit censors, including Grassi, but it was strongly recommended for publication.[29] Perhaps Scheiner's book was seen by the Jesuit scientific subcommunity as an occasion to reassert the independence of the "mixed sciences" from traditional philosophy.

The massive *Rosa ursina,* which capped Scheiner's lifelong scientific work, clearly shows that the stresses and strains in his career remained unresolved. This includes not just his evolving feud with Galileo, but also the struggles between the new sciences and the old natural philosophy, between religious obedience and intellectual conviction, between the new and the old cosmologies, between speaking for himself and for the Church or the Society. His loyalties pulled him in different and sometimes conflicting directions that he could not reconcile.

This situation was exacerbated by the publication of Galileo's *Dialogue* in 1632. This is evident in the following report of Scheiner's excited reaction to his nemesis's new book.

> Fr. Scheiner, while he was in a bookshop where there was also a certain Olivetan father [Vincenzo Renieri] who had come from Sienna during the

past days, heard the Olivetan father give merited praise to the Dialogues, celebrating them as the greatest book that had ever been published, and he [Scheiner] was completely shaken up, his face changing color and with a huge trembling of his waist and his hands, so much so that the book dealer, who recounted the story to me, marvelled at it; and furthermore told me that the said Father Scheiner had stated that he would have paid ten gold *scudi* for one of those books so as to be able to respond right away.[30]

Why was Scheiner so highly agitated? By this time his animosity towards Galileo was so strong that he may well have been disposed to argue against any new book by his old opponent. On the other hand he could no longer consistently reject Galileo's main point in day one of the *Dialogue* that the heavens are corruptible, since he himself had recently argued for that view in print. Yet Galileo also clearly favored Copernicanism despite his claim of neutrality due to the dialogue format of his new book. That violated the Church's ban of 1616, and Scheiner may have felt called upon as a priest-astronomer to refute Galileo's arguments for heliocentrism. Perhaps however his furor was raised not because he may have thought that Galileo was a heretic but because he saw him as an unrepentant thief.

This is all still quite general, however, and does not seem to be enough to account for the severity of Scheiner's reaction to Galileo's *Dialogue*. But if we look at his promised refutation, which he finished writing a little more than a year later, another reason suggests itself. Perhaps he had just realized to his horror that he himself had inadvertently supplied Galileo with the basis for one of the latter's best arguments in favor of Copernicanism,[31] and thus that he had indirectly undermined the Church's position. For a few years earlier, he had sent Galileo a copy of his recently published *Rosa ursina*, whose data on sunspots would have required many years of observation.

This requires a fuller account. Scheiner's reply was entitled *Prodromus pro sole mobili et terra stabili contra Academicum Florentinum Galilaeum a Galilaeis* (1651). As the title says, this book of 120 pages is a defense of geocentrism directed against the arguments to the contrary in Galileo's *Dialogue*. But when one examines the *Prodromus*, it is clear that Scheiner has focused almost completely on only one topic: sunspots.[32] On the other hand, except for a few isolated references to sunspots, Galileo discusses them in only two places in the *Dialogue*: three pages in day one, to argue that the heavens are alterable, and ten pages in day three, to argue for Copernicanism from the annual variations of curvature of the paths of sunspots.[33] Scheiner had already conceded the former point in his *Rosa ursina*. But the latter phenomena he had explained, as we have seen, by a cumbersome combination of simultaneous motions of the sun. For Scheiner to see that his scientific findings about curvature, based on his years of sunspot observations, had now been appropriated by Galileo to justify Copernicanism, may have been the shock that hit him in the bookshop.

We do not know precisely when and how Galileo was influenced by the curvature conclusions in Scheiner's *Rosa ursina*.[34] And certainly Galileo, not Scheiner, gets the credit for formulating the argument for Copernicanism from the variation of the curved paths of sunspots. But there is no doubt that Scheiner was convinced that Galileo had taken the empirical basis for that argument from the *Rosa ursina*, without mentioning that source, and had then misused it, in Scheiner's judgment, to argue for heliocentrism. He was furious that he had been taken advantage of by his worst enemy.

The timing of the *Prodromus* is also noteworthy. Pope Urban VIII decided to put Galileo on trial three months after Scheiner's explosion at the bookshop. The trial began six months later and ended on 22 June 1633. The *Prodromus* was completed that July. Thus during those twelve months Scheiner, who had promised to "respond right away," was actively at work writing his critique of Galileo's *Dialogue*. Hence if he in some way had a hand in instigating or influencing the course of events at the trial, his frame of mind in that context could be at least partially revealed by the *Prodromus*. However, no specific connection between these two developments are to be found.

Also Scheiner's motives in composing his "refutations" of the *Dialogue* are somewhat murky. Exactly three weeks after Galileo's trial ended, and when he was near the end of writing the *Prodromus*, Scheiner wrote the following to a Jesuit friend.

> When the *Prodromus* is finished, I will, with the help of God, defend the common astronomy against Galileo throughout the whole book, as has been recommended by the Pope, our General, and the Assistants, all in the pursuit of higher things.[35]

It may well be that these authorities had "commissioned" Scheiner to write a scientific refutation of Copernicanism and Inchofer to write a scriptural refutation. If so, the latter followed the directive explicitly, but Scheiner took cover by writing directly about his ongoing dispute with Galileo over sunspots rather than over Copernicanism. Although Inchofer refers by name to Scheiner as a scientific authority both in his 19 April 1633 report to the Holy Office on the

Dialogue and in his *Tractatus syllepticus,* Scheiner seems to have simply ignored Inchofer's contributions.

Two months after Scheiner had finished the *Prodromus,* he was described as supporting the views of Aristotle "only because of force and obedience."[36] What are we to make of this? Again the question that arises is for whom was Scheiner really speaking, himself or his religious superiors? That question is unanswerable today. But the fact that it arises is a sign that Scheiner was indeed torn between the results of his scientific work and his priestly religious commitments.

This is also supported by the curious facts that permission to publish the *Prodromus* was denied by Scheiner's Jesuit superiors, and that the book did not appear until 1651, one year after his death. My search in the Jesuit archives for documentary evidence of the specific reasons for this eighteen-year delay was not successful.[37] Could it have been that by this time some of the Jesuit authorities were already seriously concerned that maybe Galileo was right? That would indeed have been unsayable.

The Final Journey

Towards the end of 1633 Scheiner was transferred from Rome back to Bohemia, where he spent the last years of his life. His scientific work was completed. His protracted controversy with Galileo had also finally ended. On his long and wintry journey north he must have reflected in depth about the events of the previous quarter century, but we have no information about his ruminations.

Yet it is virtually certain that the central issues of these years must have remained unresolved in his mind. He was still faced with the dilemma of reconciling his pursuit of scientific truth with the demands imposed on him by his religious faith and his Jesuit vow of obedience. As we have seen, this dilemma took many forms, which in turn forced him to play many roles which were mutually antagonistic. At the personal level he appears to us to be an enigma, since we often do not know from his words whether he, in the role of Apelles again, was still hiding behind a disguise, or perhaps at times even behind a deception. We are not saying that he was dishonest; rather the circumstances of his life leave us unable to evaluate him with much confidence. In our image of him slowly trudging through the snow on his final journey northward, he is the classic case of the clash between science and religion at the personal level.

But we should not expect that he could have resolved these conflicts on his own. For in many ways Scheiner's journey still continues unabated into the present day. Unfortunately the dialogue between science and religion is still located at the same level at which he experienced it: do science and religion unavoidably come into conflict or can they be taken as two consistent and complementary phases of one, richer worldview? Although the dominating degree of cultural power has shifted from religion to science since Scheiner's day, the terrain generated by the earthquake of the Galileo affair has not really changed that much. And certainly by our day too little has happened at the more creative level of evolving a cultural synthesis of science and religion, a project that may have at least started to get underway, perhaps in the hands of the Jesuit astronomers, if the moment had not been lost.

Fallibilism and Religion

OUR GLIMPSE BEHIND THE SCENES AT GALILEO'S TRIAL HAS REVEALED
some new perspectives while still leaving many questions unanswered. Some of
the Church officials who conducted the trial seem to have been more hesitant,
and others more adamant, about the course of events than previously thought.
If my interpretation of the extrajudicial dealings with Galileo, including the
possibility of a plea bargain, be granted at least for purposes of discussion, then
the prosecutor Maculano, Cardinal Francesco Barberini, and perhaps others,
were seriously doubtful and concerned about the course of the trial. We do not
know whether this was due to worries about the legal status of the injunction
memo vs. Bellarmine's significantly less forceful letter, to a prescient feeling that
the Church would not be well served in the long run by declaring an astronomical theory to be heretical, or to some other factor. But we do know that the
issue could have been handled quite differently so that a "Galileo affair" would
not have occurred in 1633. Whether a similar condemnation on some other scientific issue (e.g., evolution) could have or would have occurred later we will
never know. We will discuss the variables entering into that situation later in
this chapter.

On the other hand it is quite clear that some members of the Holy Office were very determined to obtain a condemnation against Galileo. But unless some further documents on the matter surface in the years to come, we will not know who they were or what their motivations were. We can only speculate that they saw the impending condemnation as somehow protecting and serving the good of the church or as carrying out the wishes of Urban VIII without qualifications or hesitations. Whatever the motives, they must have knowingly taken steps to sabotage the plea bargain and then to compile the deceptive summary report, on which they knew the final decision would hinge. Needless to say the demands of justice became subservient to personal and institutional interests. And others besides Galileo were certainly affected by the repressive atmosphere of that day. For example, the Jesuit rules on orthodoxy and Christopher Scheiner's tragic personal story are full of instances of institutionally imposed silence and censorship, restrictions on intellectual freedom, demands for obedience, pseudonyms, concealment of what one actually thinks, and personal animosities. Unfortunately we have seen too many such embarrassing and unflattering pictures.

One important person of whom we did not get a glimpse behind the scenes is Pope Urban VIII, even though he must have been there all along. We do know that for more than twenty years before the trial he honored his fellow Tuscan Galileo with high praise and regarded him as a personal friend, having had as many as six private visits with him in 1624 after he had become pope. We also know that in September of 1632 his friendship was permanently transformed into a strong personal animosity towards Galileo. We do not know what specifically caused this to happen, but the most reasonable guess is that he felt that Galileo had betrayed his friendship by failing to inform him of the Holy Office memo about the injunction against Galileo in 1616. That memo had just been rediscovered in the files (by an investigating committee that included Melchior Inchofer) and brought to the pope's attention. From that point on, his anger towards Galileo only increased.

After that we have no specific information about the pope's influence, if any, on the trial. He must at least have been kept informed on these matters by his nephew, Cardinal Francesco Barberini, who supported Galileo throughout the episode, and who had collaborated with the prosecutor Maculano to settle the case by means of a plea bargain. But that is as far as one can speculate. There is no evidence that Urban VIII was involved in sabotaging the plea bargain, or in ordering or arranging the adulteration of the inaccurate summary report to obtain a condemnation. One would think that such intervention in the course of the trial could have been made only by one or more persons who were located very high up in the hierarchy of the Holy Office or by the pope himself. Did the cardinal nephew attempt to win an approval of the plea bargain from the pope, who was the one person with the unchallengeable authority to do so, or did he act on his own? Again we have no evidence at all that Urban VIII was involved, and only additional new documentation could answer that question. How Urban VIII's strong early friendship turned into his later animosity and whether he had a hand in determining the outcome of the trial thus remain mysteries. Another mystery is what Galileo had in mind at the first session of his trial when he said to the inquisitor, when he had been asked about the injunction for the first time, that Bellarmine had told him something for the pope's ear before anyone else's. What could that have been? Or was it just a bluff, or a reminder that Galileo had friends in high places?

Our glance behind the scenes shows also that some Jesuits were strongly opposed to Galileo on both scientific and religious grounds, that others agreed with him on the scientific issue (but had to be more hesitant and mute), and that the Jesuits were not organized as a group against Galileo. Christopher Scheiner, their best scientist, who was brought to Rome to refute Galileo, strongly criticized him at great length, but used the occasion primarily to highlight his own scientific work and to give vent to his bitter grudges against Galileo regarding priority disputes over issues in astronomy, especially on the discovery of sunspots. One wonders whether the excessive length and bitterness of his writings just before the trial were intended to create a diversion for him to avoid the central topic of the truth of Copernicanism, which he may have personally accepted as true.

Finally Melchior Inchofer was by far the most important Jesuit behind the scenes. He had wormed his way into center stage with his strongly negative assessment of Galileo's *Dialogue,* which was presented at the trial. He thereby ingratiated himself to the members of the Holy Office to the point that he was asked later to write a theological judgment of the Galileo verdict even before the trial had ended. Yet Inchofer's personality was so abrasive, and his ambitions so focused on the corridors of power, that in time he fell out with his fellow Jesuits. Although he quoted Scheiner several times by name to support his attacks on Galileo, Scheiner did not return the favor.

Inchofer was an odd choice for the key role he was given of making the theological case against Galileo. Nevertheless in his *Tractatus* he managed to zero in rather perceptively on the fundamental issue in the final sentence. The question he asked himself was why is it that Copernicanism is a heresy. The short answer is: because it is contrary to scripture. But to establish that it is a heresy, one specifically needs to show not only that Copernicanism is false but also that it is false in a special way. That is, its falsity must be due to the fact that it contradicts absolutely certain truths revealed by the Holy Spirit in the Scripture. So Copernicanism must be thought of as being not just false, but heretically false. That subtle point can be seen in the way that Galileo's final sentence was worded.

THE JUDGMENT AGAINST GALILEO:
FALLIBLE VS. NON-FALLIBLE TRUTH

Galileo's trial ended on 22 June 1633, with a two-part session of the court. First the sentence was read. Then Galileo had to read an abjuration statement prepared for him by the court in which he swore an oath to abandon his views on Copernicanism. The latter is the famous scene that we discussed in the first pages of chapter 1.

The sentence begins with a four-page review of the findings at the trial. This review is clearly based on the misleading summary report. The key documents used at the trial apparently were not appended to the sentence. No mention is made of the extrajudicial discussions with Galileo, nor of any possible plea bargain. Bellarmine's letter to Galileo (26 May 1616), on which Galileo had based his defense, is only partially quoted, and then to support the disputed injunction (which is not even mentioned in the letter), and thus is interpreted as damaging to Galileo. The reason given is that, "The said certificate [Bellarmine's letter] which you produced in your defense aggravates your case still further since, while it says that the said opinion [Copernicanism] is contrary to Holy Scripture, yet you dared to treat of it, defend it, and show it as probable" (Finocchiaro 1989, 290). The sentence document culminates with an explicit statement of the court's judgments against Galileo as follows:

> We [the cardinals of the Congregation of the Holy Office] say, pronounce, sentence, and declare that you, the above-mentioned Galileo, because of the things deduced in the trial and confessed by you as above, have rendered yourself according to this Holy Office vehemently suspected of heresy, namely, of having held and believed a doctrine which is false and contrary to the divine and Holy Scripture: that the sun is the center of the world and does not move from east to west, and the earth moves and is

not the center of the world, and that one may hold and defend as probable
an opinion after it has been declared and defined contrary to the Holy
Scripture (Finocchiaro 1989, 291)

The wording of this carefully crafted statement in the final judgment of the
trial is very revealing. Galileo is judged to be "vehemently suspected of heresy"
on three points. The first is the claim that the sun is motionless at the center
of the world. In 1616 the theological Consultors to the Holy Office also said in
their *internal* report that this claim is heretical, but the word "heresy" was *not*
used in the public Decree of 5 March 1616; instead it was said there to be
"false and contrary to the Holy Scripture."

The second point, that the earth is in motion and not at the center, is also
called a "heresy," while in 1616 it was described as "at least erroneous in faith"
by the Consultors, and as "false and contrary to the Holy Scripture" by the Decree. So in the official legal language used in the first two judgments against Galileo, we find an upgrade in terminology: Copernicanism is now explicitly called
a heresy.

The third point is an initially surprising element in the sentence. Galileo
seems to be judged guilty of a methodological heresy, that is, one cannot say that
something that is contrary to the scripture is "probable." Note that this third
point does not talk about any type of factual claim about the physical or the
spiritual world. Nor does it speak about any normative claim about the moral
or religious life. Nor does it say anything about "matters of faith and morals."
Rather it speaks about a relationship concerning truth values between two cog-
nitive realms: natural human knowledge and the scripture. The central topic
is the relationship of contrariety between human knowledge and the revealed
truths of scripture, and the point made is that what is identified by the Church
to be contrary to the scripture cannot be "probable." Presumably this means that
such a claim would also be a threat to the faith.

This third charge of heresy against Galileo has received rather little attention in the literature, but may well be the heart of the matter. For in the last
analysis it really makes very little, if any, difference to the religious interests
of the church whether the sun revolves around the earth, or vice versa. But
what is of the most fundamental concern to the church is that the word of God
in Scripture is the guarantor of all the religious truths of which the church
is the custodian. If that standard of determining truth were not upheld, then
Urban VIII could well have been correct when he said that the Galileo case
would be the greatest scandal of Christendom. The thrust of the third charge

is that Galileo's book is a threat to the faith because it violates the *de dicto* truth standard, as emphasized by Bellarmine, namely, that the scriptures are true simply because they were dictated by the Holy Spirit.

Given this standard to determine what is true, it follows that any religious belief that survives this test is not only true, but true with full certitude. The reason of course is that the Holy Spirit, the source of what has been revealed, is Truth itself, and cannot be the source of any error. As a result the content of the revelation is not open to the possibility of error. The same status of full certitude inspired by the Holy Spirit had also been extended to "tradition" on matters of faith and morals at the Council of Trent in 1546. However the many religious beliefs accumulated from sources other than scripture and tradition over the centuries were understood to be subject to the possibility of error. We thus find that the total set of religious beliefs falls into two categories: those that are fallible and those that are not.

The term "fallible" here refers to something quite different than the actual truth status of a claim. It does not mean that a given statement is true or false or indeterminate. It merely means that a given statement (affirmative or negative) is open to the possibility of being in error. Thus a fallible claim (while still remaining fallible) may have its truth status changed after a future consideration of new or old facts. But for a non-fallible claim based on divine revelation such a future reconsideration is not relevant, since it is already known to be true with full certitude. For example, we now know that Newton's laws are fallible, after three centuries in which they were considered to be certain, for we have discovered that at very high velocities they need to be modified. On the other hand the religious notion of a life after death, since it is part of the Creed, is taken by the church to be non-fallible, or not open to error.

Galileo was judged to be a heretic because he thought that (1) the sun is at rest in the center of the world and that (2) the earth is in motion and is not at the center. Are these fallible or non-fallible claims? The church took these two claims attributed to Galileo to be non-fallibly false, because (3) they have been declared to be contrary to the scripture, and *one cannot say that what is contrary to scripture is probable.* So the third point in the judgment against Galileo was not merely a methodological point but was essential to make the first two charges heretical, that is, non-fallibly false. If they were merely false, they would not be heresies. On the other hand Galileo must have taken the first two points to be fallible but true claims about the structure of the solar system. He always hoped to find a strict demonstrative proof of Copernicanism but failed in that attempt. And so he would likely have agreed with the view of scientists and philosophers who say today that scientific laws are always subject to the possibility of falsification of their formulation as now stated.

But that is not true of the church's religious beliefs. Some are fallible, but many are not. To use Quine's imagery from another context, the full field of religious beliefs in the church is an enormous collection of claims within which there is located, near the center as it were, a circle of absolutely certain beliefs that are inviolable and non-fallible (Quine 1961). This situation requires a clear distinction between the inner primary circle of certainly true beliefs (e.g., the articles of the Creed, the divinity of Christ, life after death) and the fallible beliefs (e.g., limbo, the immorality of usury, Christ's siblings, the superiority of celibacy) that surround the center circle. What non-fallible beliefs reside within the center circle and what fallible beliefs surround the former are questions to be answered by the church, although it would seem that the former are much smaller in number but larger in significance than the latter.

If one were to say that all religious beliefs are within the circle, that is, are non-fallible, then the result would be a very rigid, fundamentalist version of the church, where belief could not change over time, except perhaps by addition from further revelation. On the other hand if one were to say that all religious beliefs fall outside the circle, that all are fallible, then it would seem that the church has no special criterion of truth for scripture, and that the Holy Spirit is not its source for it is merely a set of uninspired historical documents. Lastly if one were to say that part of the church's religious beliefs is fallible, and another part is absolutely certain, then the church should be able to show clearly to its members where the line of demarcation lies and how it is defined. This third alternative is where the Catholic Church itself has long been located, a position that carries with it some special consequences.

For example, it would seem that the central core of beliefs can never grow smaller, since that would make some non-fallible beliefs subject to change. But it could get larger, for fallible beliefs could later be seen to be absolutely certain. (How that could happen would require detailed discussions. For example, how did the designation of "tradition," as another source of divine revelation recognized by the Catholic Church at the Council of Trent, integrate with the scripture as a non-fallible source; and why are the private personal experiences of God by mystics who are recognized as authentic not so recognized?) As the central core of beliefs increases, the relations between the component non-fallible beliefs becomes more complex, and it becomes more difficult to maintain a unified and consistent religious, moral, and physical view. Is this part of what happened at Galileo's trial? If Copernicanism is heretical, did its opposite (i.e., some form of geocentrism) then become a non-fallible belief? The sentence at Galileo's trial can be, and has been, read that way. (See Beretta [1999], and the rejoinder by Nevue and Mayaud [2002].)

Why not opt for a complete fallibilism in religion, as is generally thought to be the case in science? The traditionalist might object, "This is an unacceptable denial of absolute certitude; there would then be no absolute truth." To this the fallibilist might reply, "Of course there still is absolute truth. We even know where it is: it is in the absolute mind, but our minds are finite and not omniscient." To which the traditionalist might say, "That is precisely why God was so good to us in giving us his words spoken in scripture for our edification and salvation." To which the fallibilist might respond, "But the minds of the hearers are finite and fallible, and so can only partially understand those words." The question comes down to the problem of how a fallible human mind can grasp the fullness of absolute truth, including its non-fallible certitude, contained in a divine message. If our understanding is only partial, how can one know that the claim is not fallible?

To continue the analogy of the circle for another moment, how wide is the line that separates fallible from non-fallible religious beliefs? As we have seen earlier, Melchior Inchofer gave that line some degree of width. For he found it necessary to locate some beliefs in that middle ground precisely because he saw them as not explicitly stated in scripture, but still deduced from what is stated there by means of complex inferences, which may hide the possibility of error. It is significant that he assigned to that middle ground the claim that "the earth is at rest at the center of the world" (the denial of which was the second charge against Galileo) as a "rather probable *de fide*" truth.

I propose no answers to the above questions. But one can suggest procedural policies regarding how to live in the church under these circumstances. Given that some religious beliefs in the church are non-fallible and others are fallible, is it wise to follow a policy of increasing the non-fallible content of faith? If the objective is to increase control within the church and to remove the burden of making some decisions from the individual person, then the non-fallible teaching should be increased. But if the objective is to avoid the consequences of centralization in the Church and to give more moral responsibility to the individual person, then one should avoid increasing the non-fallible teaching in the church. Should there be a "presumption in favor of the fallible" so that the church does not become a more closed institution? Is that the lesson that Inchofer's quandary teaches us?

The Judgment against Galileo:
The Historical Consequences

If such a presumption had prevailed in the Galileo affair, the church would have been well served. For it had a chance to engage the newly born sciences in a creative interaction, as happened once before in the thirteenth century. But as it turned out, that opportunity was lost when in Galileo's day the church stopped that interaction even though it had already begun to occur among some of the Jesuit scientists of that era.

This point is worth fuller elaboration. The seventeenth century failed to bring about a cultural integration of science and religion, a condition that continues to our own day. If it seems as though we are asking too much of that era, and perhaps we are, we are reminded on the other hand that precisely that did occur a few centuries earlier when Western Christianity was shocked by the arrival of Aristotelian science and philosophy, which clashed quite directly with some fundamental Christian beliefs. For example, Aristotle presented very good reasons and arguments to show that the material world is eternal, that the human soul is not immortal, and that an infinite God can exercise no providential care and control over events in the physical world, including human life. These challenges to the religious tradition were certainly at a more fundamental level than the question of whether the earth or the sun is at the center of the world.

It would indeed have been quite possible in the thirteenth century for Aristotelian science and the Christian religion to have entered into a long term adversarial relationship similar to what has happened in modern times since Galileo. But that did not occur. Rather in that earlier era the cultural challenge of Aristotelianism elicited a robust and large-scale intellectual enterprise that ultimately integrated the religious and scientific cultures of that day into a new theological-philosophical synthesis, which was best exemplified in the writings of Thomas Aquinas. This Thomistic synthesis was so thoroughly crafted, and has had such a long history by now, that it is easy for us to fail to appreciate the depths of the chasm between the original Aristotelian worldview and the religious tradition up to that time. The resolution of this conflict required much foundational rethinking of *both* Aristotle's ideas and the then current theological understanding, as well as an openness to modify the meaning of *both* of these traditions in some fundamentally new ways. But this is precisely what did not happen four centuries later in Galileo's day. Why? The reasons lie within Galileo's trial.

Galileo's Copernican challenge to the traditional worldview involved, at a quite fundamental level, a new set of scientific methods of determining truth. He succeeded in giving only a partial delineation of these new methods in his reflective comments about science, but his own scientific work and rhetoric made them clear enough to be noticeable and challenging to others. What we would now call scientific rationality—the modern sense of an appeal to sensory evidence, to mathematical analysis, and to logical criticism in the evaluation of verifiable hypotheses—was just beginning to emerge and to be seen as distinctive as a method of authorizing what is true.

This could not fail in time to become a threat to the church's much older method of determining what is true, namely, by appealing to the "author" of the revealed words in the scriptures as the guarantor of their truth. This authorship mode of rationality is so old that it is embedded in the very etymology of the word "authority." If this mode of determining truth were to become challenged, the threat to religion would be as foundational as one could get, since the entire set of religious beliefs, as well as the complex of institutions based upon them, are grounded in the acceptance of this authority.

This is what the Catholic Church was instinctively worried about. It had to protect its religious authority at all costs for its own survival and well-being. At this basic level it was not a question about the sun and the earth. It was a question of the credibility of Catholicism. If Copernicanism had not come along to put pressure on this foundational issue of authority in the Catholic Church, some other major scientific challenge (e.g., evolution) could in time have had the same or a very similar effect.

Seen in this context the primary focus of the Galileo trial was thus the issue of authority. This fits the trial documents perfectly; they constantly refer to orders, decrees, obedience, intentions, injunctions, permissions, sanctions, and so on. The overwhelming concern was with authority and power, not astronomy or theology. And the Galileo case has continued to this day to be of perennial interest because, even after the issue of Copernicanism has long been settled, the problem of the interaction between the authority of scientific reason and the authority of religious revelation has lived on, as science and religion have remained major cultural forces. The credibility of religious authority is what the trial was, and still is, about.

The question of permanent importance that emerges from all this is whether in principle these two authorities can be reconciled as consistent and complementary, or whether they are intrinsically fated to conflict. That question has in no way been answered by our analysis of Galileo's trial. However in that particular case they certainly did conflict.

Much of the difficulty arose not directly from the nature of religious authority itself, but from how it had become institutionalized in the Catholic Church. Influenced for obvious historical reasons by the Roman imperial, and later monarchical, models of organization, religious authority had become more and more centralized in the church up to Galileo's day, and has continued since then to evolve in the same direction. By late medieval times that centralized authority also had come to exercise a hegemony in Western culture, which was broken in Northern Europe only by the Reformation. But it continued unabated in Galileo's Italy, reinforced by the highly defensive mind-set of the Catholic Church's counterreformation efforts. In other times and in other places religious authority in the Catholic Church would likely have become institutionalized along a different path. The church could have been organized differently, but as a matter of historical fact it was not.

As a result the authority behind scripturally based religion, at least in the Catholic tradition, became highly centralized, monolithic, esoteric, non-fallible, resistant to change, and self-protective. These characteristics are clearly visible in Galileo's trial in which the Holy Office operated within a pyramid of secrecy. This provided the opportunity for some members of the Holy Office to falsify documents and to prepare a quite misleading summary document of the trial, which was used as the basis to come to a verdict. And all of this was, of course, totally unknown to Galileo and his defenders.

Meanwhile in the centuries since Galileo's day, modern science has also become thoroughly institutionalized, but in a quite different fashion. By contrast modern science as an institution is pluralistic, democratic, public, fallibilistic, innovative, and self-corrective. Before new developments are incorporated into science, they must be publicized, usually in journals and newsletters; thereby they can be verified, in which case they are tentatively upheld, or falsified, leading to rejection or modification. And all of this occurs in an open and public discussion in which scientific judgments are taken to be fallible.

If it be granted that science and religion have become institutionalized in these two quite different ways, then in cases of conflict the reconciliation of the truths of science and religion becomes very difficult at the concrete level. The oldest, and still strongest, argument that science and religion are fundamentally in agreement is that two truths cannot be inconsistent, or to use the version of this argument favored by Galileo, an all-truthful God is the author of both the book of revelation and the book of nature, and thus we can be confident at the abstract level that these two books cannot conflict. But in concrete cases the institutional demands of science and religion also come into play, and agreement is thereby made more difficult. This situation is not helped by the fact that theologians and scientists are educated, especially at the graduate level, in almost total isolation, if not actual disregard, of each others' disciplines. It is then not surprising that the two mind-sets thus generated do not communicate well later when disputes arise.

This situation is complicated further by two other developments. First, in the centuries since the time of Galileo the relative cultural power of science and religion have more or less become reversed. In the early seventeenth century, religion was the overwhelmingly dominant cultural force in the western world, and what we now call modern science was a recent arrival on the scene that was not fully understood yet even by its own advocates, including Galileo. Today science and technology dominate western culture and religion is much less influential. Under such circumstances religion is understandably in a very defensive role in relation to science and technology where, for good or for ill, the big changes in society often originate.

Second, in the centuries since Galileo both science and religion have strengthened rather that weakened their institutional characteristics, which we mentioned above. For example, the definition of papal infallibility at the First Vatican Council in 1870 was clearly a further centralization of religious authority. At almost the same time the American philosopher Charles S. Peirce, perhaps in reaction to the Council's use of language, coined the term "fallibilism" to characterize science as never able to establish its data, laws, or theories as absolutely certain. This does not mean, of course, that there is no truth in science; only that such truths are always open to the possibility of being seen later to be either false or in need of modification. The nature of truth in science and in religion, and the mind-set of the scientist and the religious believer, have thus come to be seen as qualitatively different.

Clearly these reflections are not to be taken historically as part of Galileo's trial. Yet they do have their ancient roots in that most unfortunate affair. This is undoubtedly the reason why we are still talking today about Galileo's struggles with his church, even though the dispute over Copernicus's sun-centered model of the universe was definitively put to rest a long time ago.

Gratia atque vale.

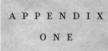

Melchior Inchofer, S.J.,
of Austria

A Summary Treatise Concerning
the Motion or Rest of the Earth and the Sun,
in which it is briefly shown what is,
and what is not, to be held as certain
according to the teachings of the Sacred
Scriptures and the Holy Fathers

Rome: Ludovicus Grignanus, 1633
with the Permission of the Superiors

Translated by Richard J. Blackwell

Muzio Vitelleschi
General of the Society of Jesus

Three theologians of the Society of Jesus, whom we have commissioned, have examined and recommended publication of *A Summary Treatise Concerning the Motion or Rest of the Earth and the Sun . . .* by Fr. Melchior Inchofer, S.J. Therefore we have granted permission for it to be published, provided that the Most Reverend Lord Vicegerent and the Most Reverend Father Master of the Sacred Palace agree. In testimony of this we present these words, written by our hand and guaranteed by our seal. Rome, 18 August 1633.

Muzio Vitelleschi

Let it be printed, provided that the Most Reverend Father Master of the Sacred Palace agrees.

A. Tornielli, Vicegerent

As commissioned by the Most Reverend Father Master of the Sacred Palace, I have seen and read with delight *A Summary Treatise Concerning the Motion or Rest of the Earth and the Sun, according to the Teachings of the Sacred Scriptures and the Holy Fathers,* whose author is Fr. Melchior Inchofer, a theologian of the Society of Jesus. The content of this book is exceedingly welcomed, especially in this day when certain people, whose ears itch because of their simple innocence, have distracted their hearing from the sacred truth of the divine Scripture and have turned to fables. This theologian has given a Christian refutation of these Pythagoreans. And he shows rightly that mathematics and the other human sciences should be subordinated to the rule of Sacred Scripture, lest in our day there occur a dangerous detour into an excessive freedom that would arbitrarily interpret, or rather adulterate, the words of God to serve the purposes of the human imagination. My judgment, therefore, is that this treatise is beneficial and should be published, especially since I have found nothing in it that is contrary to sound doctrine. Given at Rome at Ara Caeli, 22 August 1633.

Fr. Luke Wadding, O.F.M.
Commissary General of the Roman Curia

Let it be printed.

Fr. Niccolò Riccardi, O.P.
Master of the Sacred Apostolic Palace

Dedicated

To the GREAT LORD, who made the sun
and whose word speeds it on its course (Ecclesiasticus 43)
and who fixed the earth on its foundations (Psalm 103)

and

To all who love Christian and Catholic Philosophy

To the Reader

Dear Reader: In this era of the author, many hands are itching to write, and intelligent people are exhausted both by newly discovered arguments and by old remembered ones. I do not understand how there can be so much idle time for people to write about things that are forbidden and that ought to be forbidden. Such is the topic of "the motion of the earth and the rest of the sun," which has never been fully established and which has been proposed only by the Pythagoreans. In the Christian era this doctrine crept in quietly at first, but soon it was suppressed by more sound philosophy, and it survived in books in an obscure and weakened condition. Recently this opinion has been rediscovered in the realm of fictions by those who today have made it their champion, as they philosophize outside the law. But we will oppose them in favor of the decrees of the Sacred Scriptures, of the Holy Fathers, and of the theologians. Our *Summary Treatise* is only a skirmish. As a result many things remain which will quite soon be dealt with by others who are very well informed in mathematics. But if our brevity must be redeemed, we will add more to this project by taking part of our time away from other tasks, for we are especially concerned that you approve of everything in this treatise. Although we know that this work has received praise from many others, we want to rally everyone as soldiers of religion under this flag raised aloft. For everyone who has experienced the spirit of Christianity is equally obliged to act forcefully for the established faith and truth. Theologians and mathematicians and philosophers must decide to stand together and not to offend religion by having their sense of what is good destroyed by fictitious and foreign teachings. Since we wish that you, my reader, will also stand among them, we write these few words in this preface, and we have written much more in the treatise, lest you lose heart as the day grows long. Remember the words of Wisdom, "He places his honor above the figure of the scribe." Farewell.

A Summary Treatise
Concerning the Motion or Rest of the Earth and the Sun
in which it is briefly shown what is, and what is not, to be held as certain
according to the teachings of the Sacred Scriptures and the Holy Fathers

This treatise as a whole is divided into thirteen chapters. In the first, we present the main passages of Sacred Scripture that seem to support the motion of the earth. In the second, what the Holy Fathers seem to say on the same topic. In the third, what passages of Sacred Scripture indicate that the earth is at rest. In the fourth, what the Holy Fathers say on the same topic. In the fifth, we show that there are many literal senses of the same passage of Sacred Scripture. In the sixth, how many senses there are of the proposition, "The earth stands for eternity." In the seventh, which of these senses seems to be principally intended. In the eighth, that the motion of the sun is the unanimous opinion of the Holy Fathers. In the ninth, that the motion of the sun in a circle is a matter of faith. In the tenth, whether it is a matter of faith that the earth is the center of the universe. In the eleventh, that granting all this, these propositions have been affirmed up to now as matters of faith. In the twelfth, granting that the earth at rest is a matter of faith, whether and to what degree it is permissible to argue for the contrary. In the thirteenth, we present and resolve the objections that arise in regard to the Sacred Scripture and the teachings of the Holy Fathers.

CHAPTER 1

The main passages of Sacred Scripture that seem to indicate that the earth moves

There are many such passages, but most of them are to be understood figuratively. Sometimes the earth is substituted for the inhabitants of the earth. Thus the earth is said to be moved when factions among humans go to war for some reason, or better when humans change their status in life in some way, as when one generation is replaced by another. St. Augustine testifies to this in his *Commentary* on Psalm 98 [97:4]. In the same way we are to understand Psalm 96 [97:4], "His lightning lights up the world, which the earth sees and is moved," and Amos 8 [8:8], "Is this not why the earth will be moved, and all its inhabitants will mourn?" and Jeremiah 50 [50:34], "So that He may terrify the world and bring trembling to the inhabitants of Babylon," and 1 Maccabees 9 [9:13], "The earth was moved by the size of the army," although his latter passage is also explained as a hyperbole. The same must be said of many other passages in which God is said to have moved or disturbed the earth, as is clear to any one who reads St. Augustine on Psalms 17, 58, 67, 74, 113, and *Confessions,* bk. 1. Indeed there is hardly anyone else who says any more frequently than St Augustine that in Sacred Scripture the earth is substituted for its inhabitants. See books 1 and 2 of his commentary on Exodus, and his *De civitate Dei,* bk. 20, chap. 23.

Sometimes the Scripture refers to the power and majesty of God, on whose command everything clearly depends. For example, Ecclesiasticus 16 [16:18–19], "Behold the heavens and the heaven above the heavens, the abyss and the whole earth and what they contain are shaken by His gaze; the mountains and the hills tremble and shake when God looks at them," and Job 26 [26:11], "The pillars of heaven tremble and shake with fear at His command."

Finally the Scriptures refer to a miraculous disturbance in part of the earth. For example, 1 Kings 14 [1 Samuel 14:15], "The earth was shaken, and it happened as a miracle by God," or as Hebrews [12:25–27] says, "It was agitated by God," that is, it underwent an enormous and very long disturbance. In its proper sense this passage refers to the inhabitants of the earth. For as the Septuagint is translated, this passage seems to refer to stupor rather than to motion, but stupor pertains to humans. Nevertheless it is probable that the earth itself truly was miraculously shaken, especially since, as some have pointed out, the earth is often taken to mean its inhabitants, but this happens more in the prophetic than in the historical books, to which the books of [Samuel and] Kings belong. According to this view then it can also be admitted that the earth was truly moved, which is what is said at 2 Kings 22 [2 Samuel 22:8], "The earth quaked and was moved, and the foundations of the mountains were shaken and shattered, because of His anger." This is repeated in almost the same words in Psalm 17 [18:7].

From all these and other similar passages one cannot infer that the earth has either an annual or a daily rotation, which is what we are concerned with here. Nor can that be inferred from the motions and disturbances of the earth, which frequently arise from natural causes.

There is one particular passage to which some have attributed great importance: Job 9 [9:6], "He moved the earth from its place and its pillars were shaken." Diego de Zuñiga has argued that this passage alone is enough to infer the motion of the earth, because no other passage is to be found elsewhere in Scripture that shows on the contrary with equal clarity that the earth is not moved, while this passage says that it is moved. He also thought that this passage can be easily explained by the teachings of the Pythagoreans, who believed that the earth is moved by its own nature. Some also claim that this accounts for the movement of the mountains, which is mentioned in the same text [Job 9:5], "He moved the mountains, which did not know it, and He destroyed them in His anger." They claim that this motion is above all to be understood as a circular motion. Furthermore it would be extraordinarily astonishing for the earth with its heavy mountains to have a fast circular motion unless it were so constituted from the beginning of creation.

But in fact these passages prove nothing. In regard to the first one, scriptural writers have correctly pointed out that it refers not to the earth's immobility, but to its stability, since the word "pillars" implies firmness and fixity. In truth the word "motion" refers to God's punishment and the power of His aroused anger in the form of storms and convulsions on the earth, which in no way recur on a daily and regular basis, as these authors see it. They correctly deduced the same thing from the words of Hebrews, which seem to refer, not to a natural motion, but to a disruption arising from fear or anger or some other such disturbance.

Regarding the second passage, it is clearly agreed that it would be wrong to think that that motion is the same thing as the circular motion of the earth. For an action arising from anger and destruction would bring such a circular motion to an end; but such an action should continue for a long time if it is for the purpose of benefitting mortals here on earth. Hence the Scripture speaks of the motion of the mountains from place to place, which is sometimes done by God to punish people; and this has happened not just once or only in one place, as both history and human traditions teach. It certainly cannot be denied that mountains could also be moved for another purpose, that is, to increase the prayers of the saints and the faith of the people. But in either case it does not follow that cities with their inhabitants would be buried or utterly destroyed.

CHAPTER 2

What the Holy Fathers say about the motion of the earth

I have not found a single one of the Holy Fathers who has dealt with the motion of the earth clearly and positively, as the saying goes. But from some of them it is possible to deduce a few things that seem relevant here. Commenting on the words of Psalm 118 [119:90], "You established the earth as permanent," (which others interpret as referring to its immobility) St. Ambrose adds, among other things, "Ecclesiastes [1:3–4] beautifully illuminates this passage for us as speaking spiritually, not materially. For whatever abundance man acquires from all of his work under the sun, the generations come and the generations go, but the earth stands forever." Now if the meaning of these words is spiritual rather than literal, then it is not correct to infer that the earth is permanent, that is, that it does not move circularly, but at best that it does not move locally as a whole. Or rather, according to St. Ambrose's view, sublunary generation and corruption always occur on the earth, but it itself is neither generated nor corrupted as a whole.

We can refer here to Theodoretus, Haimo, and St. Bruno, who also give a moral interpretation to this passage, and to Psalm 92 [93:1], "For he has established the orb of the earth which will not move," and to Ecclesiastes 1 [1:4], "The earth stands forever," which especially seem to prove that the earth is at rest. And Dionysius the Carthusian interprets these same words of Psalm 92 as more suitably referring to the stability of the church rather than the earth.

Those who deny that the heavens are spherical would agree. For how could one ever infer from this that the earth stands motionless in the middle of the universe as a center around which the heavens revolve? How could that happen if the heavens do not have a spherical shape?

One of these seems to be St. Justin Martyr in his *Ad orthodoxos*, q. 59. The problem is how could the heavens not be spherical since the sun is concealed at night. As is clear in his reply, Justin claims that the concealment of the sun is caused by something other than the spherical and round shape of the heavens. Many agree with this opinion, especially those who are opposed to Aristotle.

Lactantius in his *De falsa sapientia,* bk. 3, chap. 24, argues against those who believe in the antipodes, and who therefore say that the heavens are round and that the earth, located in the middle of the heavens, is also like a sphere. But he replies, "I could prove with many arguments that it is in no way possible for the heavens to be below the earth." He also could easily have said that the earth does not rotate on its center, nor can it undergo any motion.

In his *De Genesi ad litteram,* bk. 2, chap. 9, St. Augustine leans toward the view that the heavens are stretched out like a tent, as is found in Psalm 105 [104:2]. He doubts whether it can be proven with certain arguments that the heavens have a spherical shape, and hence are convex, which he says "is a strong figment of the human mind." Hence it can be said that the earth is not located in the middle of the universe, and that there is no motion around its center.

Finally I could mention many others like Procopius who, in his commentary on Genesis 1, denies that the earth is a sphere or a hemisphere, which he proves from Genesis 19 [19:23], "The sun rose over the earth." He says, "The sun would not ascend, which is rightly said, if the heavens had a spherical form." He adds other passages in support from Psalm 18, Matthew 24, and Hebrews 8. These texts, he says, clearly refute the opinion of those who say, contrary to the Scripture, that the heavens are spherical.

Now if the heavens are not spherical, then the earth will not have any motion around its center. "For if there is no center, neither can there be any motion around it," as Cardinal Cusa has argued in his *De docta ignorantia,* bk. 2, chap. 11. After much discussion he thinks that he has demonstrated the motion of the earth, although he assumes many things that seem paradoxical to others, for example, that the world has neither a center nor a circumference, on which he partially agrees with Procopius. In Chapter 12 he claims that the earth in fact does move, but this is not apparent to us who perceive motion only in relation to some fixed point of comparison.

Since we are examining here only the opinions of the ancient Fathers, we will pass over those who follow the pagan philosophers mentioned by Plutarch in his *De placitis,* bk. 3, chap. 13, and who thus have revived in the Christian era the fiction that the earth moves. Among them were Francisco of Ferrara and a group led by Nicolas Copernicus, namely, Caelio Calcagnini, Andrea Cesalpino, William Gilbert the Englishman, and not a few other more recent and most recent people who maintain that the motion of the earth has been clearly demonstrated as a fact.

The fact that Scripture sometimes speaks of the orb or circle of the earth, as in Isaiah 40, Psalm 92, 95, etc., does not allow one to deduce that the heavens and the earth are spherical and round bodies. So from what has been said, some have further inferred in a parallel way that the fact that the Scripture sometimes seems to say that the earth is at rest, does not allow one to infer its absolute immobility, especially when such language can also easily be given another interpretation.

But these views are easily refuted today. First however let us note that at the time of the Fathers it had not yet been clearly demonstrated that the heavens have a spherical shape. If that had been proven, St. Augustine would have maintained that what he said about the tent is not contrary to those true proofs, even though it does disagree with things said elsewhere. For he would have said that the heavens are like a suspended vault, and the shape of a vault could consistently be spherical. So if the heavens are spherical, they are a vault. He examines many passages in this way, even to the point of reducing the meaning of "tent" and "vault" to a figure of speech. Moreover, although today it seems to have been demonstrated with sufficient clarity that the heavens are round like a globe, as is granted by all who have studied the matter, nevertheless there are still differences about what the truth is concerning the motions of the heavens and the location and other properties of the earth. Furthermore the force of this demonstration is derived primarily from the necessity that any other shape for the heavens is excluded, because the same distance is always maintained between the fixed stars as they move, because the stars are always the same distance away from our view, and because astronomical evidence and instruments would never all agree with each other if the heavens were not spherical. But all these matters which we have presupposed here need not be examined further.

CHAPTER 3

What Sacred Scripture asserts about the earth being stationary and at rest

First of all this can be deduced from those texts that attribute a state of rest and an unmoved stability to the earth. For example, Isaiah 66 [66:1], "God says

that the heavens are my throne, and the earth is my footstool," which is repeated at Acts 7 [7:49]. Now the heaven that is called God's throne is best understood to be the empyrean heaven, which is immobile, according to Clement
of Alexandria, *Stromata,* bk. 4; Albert the Great, *De quattuor coaevis,* q. 4, art. 15;
and many others. So when the earth is called His footstool, it must also be taken
to be immobile and at rest, especially since that is also said about the throne of
God. Again Isaiah 40 [40:22] says, "God sits above the circle of the earth." Many
have rightly understood this to mean that the motion of the celestial heavens
is not only around a fixed center, but is also contained within the immobile,
that is, within the empyrean, heaven, as St. Bonaventure says in [*Collationes
in Hexameron*] I, dist. 2, art. 1, q. 1. The same thing seems to be pointed out
by Pico [della Mirandola] in his *Heptaplus* as the opinion of the Jews, and before him by Dionysius Rickel [Dionysius the Carthusian] in his *Opusculum
de rerum productione,* chap. 58, and *Opusculum de creatione rerum ad Deum
consideratio theologica,* chap. 58. This shows that the Holy Fathers were referring to the empyrean heaven. And in his commentary on Genesis, art. 6,
he attributes not only to Bede and to Strabo (as many do), but also to St. Basil,
the view that the words, "In the beginning God created heaven and earth,"
refer fundamentally to two things, above and below, which remain stable,
namely, God's throne and His footstool. There is nothing to prevent the word
"throne" from being taken sometimes literally and sometimes spiritually, which
[Pierre] Galatinus [or Colonna] points out among other things in his *De arcanis [catholicae veritatis],* bk. 7, the last chapter. For it is a well-known doctrine among the Holy Fathers and scriptural exegetes that the words of Sacred Scripture admit of many meanings beyond the literal and historical sense.
We will indeed explain later that there is sometimes not just one meaning to
the words.

   There are other passages that seem to indicate the same thing even more
clearly: Psalm 74 [75:3], "I established its pillars"; Psalm 92 [93:1], "For He
established the orb of the earth which will not move"; Psalm 95 [96:10], "For
He structured the orb of the earth which will not move"; Psalm 103 [104:5],
"Who established the earth on its own foundation, and it will not move";
Psalm 118 [119:90], "He established the earth as permanent"; 1 Paralipomenon 16
[1 Chronicles 16:30], "He made the world immobile"; Isaiah 44 [44:24], "I am
the Lord who makes all things, extending the heavens and stabilizing the
earth, and no one helped me." These passages can perhaps be taken as a denial that there is any shaking or disturbance of the whole terrestrial orb, or as
a denial that it moves from place to place, but not that it has a daily or annual

revolution on its axis in the same place. Or they could signify a stable duration of the earth in which it does not undergo generation or corruption as a whole. Or finally they could reveal the immense power of God, who sustains this gigantic machine without any pillars or pedestals, as is said in Job 26 [26:7], "He suspended the earth above nothingness," and in Job 38 [38:6], "On what are its foundations based, and who has laid its cornerstone?" which is astonishing for us mortals who do not understand such weight and stability. Although I agree that these passages can be so understood, the main sense of the words is to be found by joining them on all sides with other passages of Scripture that affirm that the heavens are up and the earth down, that the sun rises and sets and circles through the meridian and returns to the north, and so forth. At the same time nothing similar is said of the earth, but rather everywhere its immobility and stability are pointed out. Also it should be considered that at creation the earth was given the primary quality of being heavier than the other elements, and thereby it has the character of being the center. Finally let it be noted that if the meaning of the "stability of the earth" is considered both in its own right and in conjunction with the Scriptures, then it must undoubtedly be concluded that every type of rotation is absent from the earth. This, among other things, is clearly indicated by the words of Isaiah 44 [44:24], "I am God who made all things, extending the heavens and fixing the earth," as if He stretched out the heavens like a field to allow for the motion of the stars, while on the contrary He stabilized the earth at rest. This is still more clearly said in Psalm 103 [104:5], "He established the earth on its foundation, and it will not move forever." Clearly if the earth has some sort of rotation, it would not rightly be called stable and fixed. And if it is permanent, then in no way can it become inclined, because there cannot be any kind of rotation without an inclination.

It is truly astonishing that, although the Scripture often speaks of the motion of the heavens and the stars, it never speaks of a similar motion happening to the earth. This latter would necessarily lead us to recognize the great power of God and to be carried away in admiration for Him, since we mortals, who would be rotated with the unperceived motion, would not experience any disturbances, since everything would turn out as well as one could hope. But if this arrangement of things is granted, then we would have to say either that the earth floats on the waters, as some of the Holy Fathers thought because of Psalm 135 [136:6], "He established the earth on the waters," or else that the earth was placed next to the waters, as other Holy Fathers thought. But this could not happen without great disturbances from such a change, unless the waters were held back by additional miracles.

If all this be granted and be carefully considered, then from various passages of Sacred Scripture, and especially from Ecclesiastes 1 [1:4], "Generations come and generations go, but the earth stands forever," the exegetes have correctly claimed not only that the duration of the earth is predetermined, but also that it has no motion of any kind. The first point is concluded from the Arabic version of the text, "The earth endures, or will endure, forever." Here the first phrase is contrasted with the second one as inconsistent; and while this indicates a certain wavering due to weakness, it signifies a constant firmness and permanence, like a pillar standing immovably in one place. Also the meaning of the word "stands" is exhibited often in Scripture, as when it is said in Joshua 10 [10:13], "The sun stood still in the middle of the heavens," and in Jeremiah 15 [15:1], "If Moses and Samuel were to stand before me," that is, as remaining fixed and unmoved before the word of God.

Another point also clearly follows. It comes first from the contrast made to the motion of the sun, when the Scripture immediately adds [Ecclesiastes 1:4–6], "The sun rises and sets and returns to its place, from where, reborn, it circles through the meridian and is reflected to the north." These words could properly be attributed in the same way to the earth, if the earth were to undergo a daily or an annual rotation. But then it could not be said that "The earth stands forever," for such a state of continuous rest is contrasted with the motion of the sun, and with no other body, which up to now our senses have seen and which reason has proven. There are those who think that these words are asserted only according to the human senses and according to appearances, such that on the contrary it is not the sun but the earth that has this motion. But these people speak rashly and bring the force of Scripture against the opinions of all the Holy Fathers, as we will soon explain. Moreover, if speaking like this were permitted, then the literal meaning of Scripture will vacillate in many ways, and faith in the things we hope for will be overcome and destroyed by human fictions.

Secondly this also clearly follows from the word "stands" itself. For this signifies that from the beginning of creation the earth was constituted in its nature to be inclined toward the lowest part of the universe by its weight and heaviness, thus being devoid of all motion and seeking only to be at rest at the center. For otherwise what reason could there ever be for it to be burdened with so much weight, if it were supposed to rotate rather than to be located in the middle so that the celestial sphere could rotate around it? A center by its very nature signifies stability, and it is clear that the original word "cen" is from the root "cun," which means to be firmly established. Of course, the center is the base, and it is at rest, devoid of all motion; which is why the Scripture in Job 34 [38:4] uses the word "foundations." This is how St. Basil interprets Isaiah 13, just as a little earlier he explained the words in the passage cited from Job [38:4], "Where were you when I established the foundations of the earth?" St. Basil says, "Contrary to the views of the secular teachings of the professors, God laid the foundations of the earth in the center of a sphere, laboriously encompassing the mass of the world, so that the earth would safely fall from the heavens equally balanced, equidistant, and equally solid on all sides; and it was firmly supported from below."

This is the reason why the ancients, and especially Plato in his *Timaeus* [55d–56a; see Cornford 1937, 210–39], represented the earth by the cube as its proper symbol. For the cube has the property that, however it be rotated, it always remains [*haereat*] unchanged, which is derived from the Hebrew word "Erets," which means "earth." The cube has the following mystical origin. Three-sided figures [isosceles right triangles] are formed by assembling pairs of points. The cube is then generated by combining these triangles. The cube is generated when (1) two sides [hypotenuses] are brought together and connected [a square], (2) the other sides are considered to be unconnected, and (3) this process is repeated six times. Thus sets of three points are connected one by one to form triangles, which when added together form a cube. Any such figure considered geometrically is the least susceptible to the motion of bodies, which it completely resists.

What is true of numbers is also rightly found in figures. Indeed when four rectilinear triangles are joined together into one figure, a solid body which precedes the three-cornered cube is generated. This first of the solids is the pyramid, which is endowed with three angles, and which has no capacity for motion (Piero [della Francesca, *Libellus de quinque corporibus regularibus*], bk. 38). A second solid is generated by the points formed into triangles which serve as the elements of a cube. They are not easily fastened together to form what is called "earth" as constituted of three-cornered cubes, but they indicate that the immobility of the earth is a name given to the earth by nature. This perhaps is behind the hieroglyphic of a simple equilateral triangle, which the Egyptians called a sign of divinity, as if it were the same, and in their teachings they accounted for God's essence as unchanging by this figure. Furthermore from this it still follows that, since the invariable center of all created things is God, from whom all things flow forth and are gathered back in the same continuous tempo, as seems to be the view of Dionysius in his *On the Divine Names,* chap. 5, so also the earth was established at creation to be the center of the material universe (which we will prove at the proper place), and by this symbol and its similarity the immobility of the earth is correctly, if sketchily, represented. Just as the ancients represented the earth as a cube, they also by contrast attributed the icosahedron to water, whose excessively fluid mobility in every direction is diametrically opposed to the immobility of the earth.

A third thing can be concluded from the same original word for "earth" which was derived from the Hebrew. In the Scriptures this word often signifies primary matter, which is denoted in the Scriptures by no other name. Indeed many distinguished theologians and Holy Fathers have, not without merit, taken the word "earth" to mean simply raw and unformed matter. This is clear from St. Augustine's *Confessions*, bk. 12, chaps. 20–21. Furthermore, although primary matter is the underlying subject of change in all things open to generation, nevertheless it itself is subject to no alteration nor to any change of any kind. So "earth" correctly designates primary matter in this sense. On the other hand the secondary matter of all things open to generation undergoes local motion as a whole much more than any kind of alteration. This is also the reason why all the things that are on the earth press down upon it with their weight and, as another root in Hebrew words indicates, they push and rub against the earth, yet it itself remains immobile. Moreover if primary matter is not signified in Scripture by virtually any other word than "earth," this is because not all things are expressed by their appropriate names in the Hebrew language. In his second *Homily on the Hexameron* St. Basil makes this same complaint when he says that the word "earth" designates matter in Genesis 1 and Ecclesiastes 1.

C H A P T E R   4

What the Holy Fathers thought about the state of rest and immobility of the earth

Regarding the Holy Fathers it must be noted that they presupposed, rather than argued, that the earth is at rest, in agreement with the common opinion of the philosophers. Thus in his commentary on Isaiah 13 Basil agrees that the earth is the center of the world, as that is the opinion of the professors of a different science. Also in their examination of the more interesting passages in Scripture, the Fathers did not devote much attention to questions involving mathematical disputes. Thus in his first *Homily on the Hexameron* Basil does not wish to examine various things, including particular properties of the earth. He says, "Let us also add that such studies and efforts contribute nothing of value to the edification of the Church." And a little later, "With these words we give the same advice concerning the earth itself, so that you do not inquire into its substance as being more interesting, and so that you do not weaken the liveliness of, and uselessly waste away, your powers of thinking by inquiring into that topic." In regard to our topic of the motion of the earth, it is quite possible that some recent thinkers have distorted their power of reasoning when they try, with laborious but empty imaginations, to revive the rejected view of the ancient Pythagoreans, as if they had found something new and unheard of. Similar advice is given by St. Ambrose in his *Commentary* on Psalm 118 [119:90] regarding the words, "He gave the earth a foundation which is permanent"; by St. Augustine in his *De Genesi at litteram,* bk. 2, chaps. 9–10; and by Isidore of Pelusium in his letter 273 to Ophelius Grammaticus, and in letter 100 to Paul. From their frequent remarks on this, we can conclude that the advice of the spiritual Holy Fathers, as Ambrose says, is to gleefully know and to teach the future things of eternal life, and not to be concerned at all with the more subtle content of mathematics. Likewise Isidore of Pelusium recommends the exploration of all things which relate to living well and beautifully, which are difficult to see. In the work cited above, chap. 9, St. Augustine says, "Many have argued at length about these matters, but with greater wisdom our major writers have omitted them. For they are of no value for those who seek the blessed life, and what is worse, they consume far too much time which should be devoted to more beneficial things. What is it to me whether the heavens are like a sphere that includes the earth, which is a mass suspended in the middle of the universe, or whether the heavens are like a disc that covers the earth above one side?" And Isidore of Pelusium says, "Virtue is not increased when someone undertakes an inquiry into the path of the sun, or the sphere of the moon, or the shape and size of the stars and the earth; the same is true of all other topics, when men, who think they are wise and learned and who investigate what is curious, slip away from the truth and waste their lives on trivia and foolish words."

Although the Fathers presupposed that the earth was at rest rather than inquire beyond what was established, still when the issue did arise, they did not neglect their duty. For example, in the passage cited above in his first *Homily on the Hexameron,* St. Basil does not merely presume that the earth is immobile, but he appeals to the opinions of others, and especially of Aristotle whose arguments he uses, to show in many ways that the earth is the center of the universe. And then he adds, "Our discussion has proven that the center of the universe is located within the earth." And a little later he says, "If what has been said seems credible to you, then transfer your admiration to the wisdom of God which has so arranged and disposed all things. For the admiration that arises from the greatest things should not be diminished because a method has been found that reveals what is admirable. But if this is not the case, then surely the simplicity of the faith takes hold of you more strongly than rational proofs. The same will be true after we will have spoken of the nature of the heavens." Thus it is that the great Basil has judged the teaching of Aristotle on this point from afar and with more good sense than some more recent thinkers. He compresses this same teaching in fewer words in his commentary on Isaiah 13, and elsewhere in passing, although he does not seem to have explicitly intended it.

Although St. Ambrose says in the text cited above that it was not a concern of the Holy Fathers to describe the axis of the heavens or the distribution of the elements or many other topics in a philosophical way, still when he does consider such matters, he does not deny, but rather affirms, that the earth is in the middle and that it has the role of being an immovable center. Specifically he says, "It does not have its stability because it is suspended in the middle by some lance and remains stable because of its weight. Rather this is because the majesty of God freely constrains it by a law that holds it motionless above an unstable emptiness. It does not rest on its own foundations or supports, but rather because God has ordered that it be held by the support of His own will, for all the boundaries of the earth are in the hands of God." He says the same thing more than once in his *Hexameron*, bk. 1, chap. 6. The same view is found in Basil's first *Homily on the Hexameron*; in Gregory Nazianzenus's *Orationes*, 34; in Chrysostom's comments on Psalm 102 [102:25], "He established the earth upon its stability"; in Hilary's comments on Psalm 135 [136:6], "He established the earth above the waters"; and in Damascene, [*De fide orthodoxa*,] bk. 2, chap. 10. Nevertheless they do not exclude natural causes, but rather they trace the admirable reason of this stability back to the first cause of things, which is the will and power of the creator. Thus Ambrose says in the place mentioned above, "Others marvel at the fact that the earth never departs from the central place, which it occupies by its own nature, and that thus it must remain in its place and not deflect in another direction when it is moved according to its nature and not contrary to its nature, etc. However I am unable to comprehend the depths of His majesty and the excellence of His skill, and so I do not undertake any debates about weights and measures. Rather

I consider that all things reside in His power, and that His will is the foundation of the universe, and because of Him the world remains still here." So says Ambrose, who nevertheless in the same place discusses the natural motions of the elements, and rightly states that no one motion is appropriate for all of them. Finally, up to this point in his *Commentary* on Psalm 118 [119], and before he will have said that it is not the concern of the Holy Fathers to philosophically discuss the axis of the heavens and the location of the elements, he had already maintained that the earth is in the middle as a center. He says, "The earth is the foundation on which we stand, it being granted that it is in the center of the heavens. Outsiders praise this view of the Aristotelians, and Scripture seems to say the same thing in the words of Job 26 [26:7] that the earth is suspended in nothingness. Hence the earth is enclosed in the celestial sphere, and thus the sun is not visible at night because by rotating it is located then in the lower part of the celestial sphere." If the words of Ambrose in this passage are carefully examined, they testify that the opinion of the philosophers who locate the earth in the center seems to agree with the Sacred Scriptures.

In his *De Genesi ad litteram,* bk. 2, chap. 9, St. Augustine says that the shape of the heavens, etc., is no concern of his, but still he adds, "But the credibility of the Scriptures is involved here. And for that reason I have often pointed out that someone who does not understand divine revelation, and who has found or heard something about such topics from our books which seems to disagree with what he already takes to be true, will not believe other useful admonitions in our books. It must be said briefly that our authors knew the truth about the shape of the heavens, but the spirit of God who spoke through them did not choose to teach men these things which are of no use for salvation." So said Augustine, who thereby accounted for many views that seem to be contrary in their subject matter. If his words are accommodated to our topic, we can say that our authors (and certainly the Apostles and the ancient Fathers who interpreted the Scriptures) knew what is true about the immobility of the earth, but the spirit of God who spoke through them did not choose to teach men these things which are of no use for salvation. Nevertheless at times there will arise, and today there are many, people who are unstable, who distort the Scriptures, and who do not acquiesce in its words and teachings with piety. Rather than upholding sound doctrine, they surround themselves with teachers who meet their desires, and with itching ears they turn away from truth and to hearsay. They become converted to fables and falsely mix in with them a new system of the heavens. Now these people affect the credibility of the

Scriptures, for if what is not useful is not reconquered, it will also bring down what is useful. And since the Pythagoreans have gradually come to oppose the faith, it must be shown that the truth that is found in the Scriptures, and that our major authors knew, is opposed to them. Furthermore in regard to the other question of the motion of the heavens, Augustine responded in the same way, by refuting the arguments of the opponents, and by always trying to leave the objective truth of the matter uncondemned, even by allowing various hypotheses.

Isidore of Pelusium does not think differently. And if there are others who would say that this argument does not apply to them, they have chosen to simply bypass the question of the immobility of the earth rather than to call into doubt with empty inquiries something that they accept as certain. Isidore himself seems to have done this, since he lists some of the properties of the earth but remains silent on its being at rest, as if he presupposed that as certain. Further if we were to choose to take into account others who expressly affirm this, or implicitly indicate it, or treat as certain something else from which it necessarily follows, I fear that we would exceed the limits of this treatise. We leave those few authors aside.

In his *Exhortation to the Gentiles* Clement of Alexandria says, "He abundantly provided you with a harmonious and elegant universe, and He gathered the opposing elements into an agreeable order so that the whole world would become a harmony. He regulated the freedom of the sea to prevent it from invading the land, and on the other hand He brought stability and solidity to the earth, and predetermined the limits of the sea upon it." So said Clement. From these words it should be noted that it would hardly be correct to say that God collected all of the opposing elements into a harmonious order if the most heavy and dark earth could be located either among the planets or not in the lowest place among the elements, but held a place in the middle of the universe as a matter of chance. Indeed that harmony in which the opposition of the elements is collected into an order ought to be visible in all things, but most especially in location, which is such a necessary consequence from their primary and secondary qualities and their similarities and differences, as the Aristotelians maintain. And that is what Clement clearly means when he says, "Indeed fire even softens the force of the air, just as Lydia softened Doric speech," that is, by tempering sharp and loud sounds with more soft and quiet ones. In other words it should be said that the earth and water and each of the elements are harmonized with each other, as if they were the extreme sounds of the universe, in the same way that a composer harmonizes like Nete, that is, that the outside strings in musical instruments, which surpass the others in bassness and sharpness, are rhythmically and tastefully balanced. In this way He separated the earth from the sea and constrained the earth by a law of stability and rest; and thus the water came to splash against it. It cannot be denied that Clement's opinion is that the stability of the earth makes it opposed to motion and to the unhindered and reciprocal seething and raging force of the sea, and in like manner all the natural motions on the earth are circular, either perfect or imperfect. Nevertheless he also says that the earth was made solid to be moved. It is said of this opinion either that he taught that the earth was held up by the waters, or better that he thought that the earth was established above the waters.

In his response to question 59 [in *Ad orthodoxos*] St. Justin Martyr says, "Heaven and earth were at rest in the first heaven, because they originated in isolated places, and the earth can be compared to the heaven because they were made at the same time. But the earth cannot be considered then to be in the firmament, for the earth was established before the firmament, and its whole substance was wet, and it was not affected by the firmament, which was made after the earth." Justin speaks here of the first or empyrean heaven, which is immobile, and he compares the earth to it as also immobile. The earth originated in isolation, that is, at an extreme distance from the heaven, and thus it was held at rest at the center in the lowest place. In chapter 1 [2] we discussed the question of how the fact that the sun is certainly blocked at night could ever disprove that the heaven is a sphere. He seems to give another answer here, but one that is only probable and ought not to be taken as fully adequate. Finally he does not deny all circular motion, from which the sphericity of the heavens could be deduced, as is clear if one considers the example given in the same place, and the response he gives to the next question, which we omit here for the sake of brevity. Moreover the fact that he wished the earth to have originated in an isolated place so that it could carry a wet substance, is not enough to indicate that it was established at the lowest place and that it is not moved in any way. For Justin "carry" in this text means the same as "sustain," otherwise he could not have avoided a motion from place to place.

In his *Adversus gentes,* bk. 3, Arnobius of Africa tells us that some of the ancients, who for various reasons call the earth the Great Mother or Ceres or Vesta, thought that the earth alone is at rest in the world, while all the other parts of the world were in perpetual motion. Although their stories that the earth is a goddess are just fables, still they had noticed an old and accepted truth about the immobility of the earth, and they have greater authority on that point. And while Arnobius rejected these fables, he nevertheless did not repudiate the notion that the earth is fixed in place while the other parts of the world are perpetually moved.

In his *Oration against Idols* St. Athanasius seems to think that the earth is carried on the water, but he never explicitly says that the heaviest of all things does not naturally sit still, but rather that it is stable and immobile. In this and in other ways the earth differs from water, which, though it is heavy in itself, still flows downward, while meanwhile the earth remains immobile and is held in the center where it was originally composed. One would speak truly and correctly if one were to ask, assuming the earth were to rotate, why the parts of its circumference do not fly apart from top to bottom, and how it could be held together in the middle. Athanasius finally adds that, "The permanently immobile earth is abundantly fruitful, and gives rise to humans who in contrast live and then die."

In his commentary on Psalm 12 St. Jerome says that the earth is the lowest of the elements "because the earth is the limit of creation," which, being lowest, is thus not above the sun. And on Jeremiah 13 he says, "The Lord has said that, after the heavens will have been capable of being measured upwards and the firmament of the earth to have been explored downward, I will destroy the whole nation of Israel, etc. And just as the heaven cannot be higher than it is, and the earth cannot be lower than it is, so also the nation of Israel will in no way be capable of being rejected." If these words are true, how could the earth be located outside the center, resting among the starry planets above the sun in the center? On this the text of the Septuagint edition should be noted: "Even if the heavens will have been raised higher and the stones of the earth brought downward, I will still not reject the nation of Israel." The loftiness of the heavens is contrasted here with the lowness of the earth, as a roof to a floor, and without doubt they are at an extreme distance in the house of the universe. Hence the earth must be in the middle, motionless in every direction like a floor and a foundation. And Jerome reasons that just as it is impossible for the heavens to be carried any higher (that is, by the forces of nature), so also the earth cannot be carried any lower, and so it cannot be moved. His meaning can be seen more easily in the text of the Septuagint, from which Jerome took his syllogism that, just as it is impossible to measure and to know the height of the heavens and to explore the foundation of the earth, and to comprehend these extremes by reason, so also is it impossible for the whole nation of Israel to be rejected. If the earth were not in the middle of the universe, suspended in nothingness, and not truly unmoved, it would be difficult, if not impossible, to understand the nature of the planets, which wander on their paths through a fluid medium, and to understand the nature of the sun at rest when we know that for so long it has moved in many ways.

In his second *Homily on Genesis 1* St. Chrysostom says, "In the beginning God created heaven and earth. Note," he says, "how the dignity of God shines forth in his workmanship. For in building his house God went beyond human custom by spreading out the heavens first, and then later laid down the earth beneath it. The roof came first and the foundation was later. Who has ever seen or heard such a thing?" Indeed since the heavens are the roof and the earth the foundation, then the former is at the top and the latter is at the bottom; and just as the foundation of a house does not move, then neither does the earth. He explicitly says that same thing in his second *Homily on the Feast of the Finding of the Holy Cross,* namely, that the earth is inanimate and immobile, and thus it has no circular motion, which can occur only if there is present an intelligent or an instrumental form. Indeed, if no such soul is present, then there is no significant reason why the earth would move in one direction rather than another, which is determined by an intelligent or an instrumental soul. Finally in his *Homily 12 to the People* he says, "The earth is fixed, but the waters are always in motion, and not just the waters but also clouds and rains, which are continually and repeatedly restored at the proper time."

In chapter 44 of his *On the Nature of Things* St. Isidore of Seville says first that the position of the earth is adequately established by Job 26 [26:7], "He suspended the earth in nothingness." He then cites the opinion of the philosophers who think that the earth is surrounded by, and sustained by, dense air, such that the earth is immobile, like a sponge which is balanced on every side and which cannot incline toward any direction. Moreover he asks whether the earth is suspended on water, as is said in Psalm 135 [136:6], "He set the earth upon the waters," and how it could then maintain an equal balance so that it would not lean in some direction. Thus Isidore presupposed that the earth does not have any kind of circular motion, just as a sponge likewise does not move in the air, and as no body moves in water in which it is equally balanced. He praises Ambrose for holding this same opinion. As to how this happens, he says, "It is not permitted for any human to know, and no one is allowed to investigate or discover, how by the excellence of divine skill the majesty of God has established the law that the earth remains at rest, either above the waters or above the clouds."

In his *De fide [orthodoxa],* bk. 2, chap. 10, St. John Damascene says, "The earth is one of the four elements, endowed with dryness, coldness, and heaviness, and devoid of motion from the very first day of its birth. No human can say how it is defective in any way or on what foundation it rests." While various people have had various opinions about this, everything should be referred to the power of God, as did Ambrose, Isidore, and others. If the earth were to move around the sun as one of the celestial planets, then it would hardly be correct to say that it would rest on a foundation as it moves, and no human could say this.

In his comments on Genesis 1 [1:1], "In the beginning God created heaven and earth," Procopius gives the same interpretation that we quoted just above from Chrysostom. That is also found in his commentary on Isaiah 13 [13:13] on the words, "I will agitate the heavens above, and the earth will be moved from its place." As he himself says, "The heavens will be angry, and the earth will be moved. Those who pursue other studies do not hesitate to say that the earth is at a point in the heavens which surround it on all sides; and it must be immobile since it is suspended with an equal balance and with equal intervals of distances on all sides." And on the words of Isaiah 42 [42:5] he says, "He made the heavens and He fixed them, and He stabilized the earth and what is on it . . . ," and "Isaiah proclaimed that the heavens are fixed in order to indicate that the water stands still, as Moses thought, but not fire, as was claimed by the secular wisdom of the teachers. And he proclaimed that the earth is stabilized. Hence some think that the earth remains immobile since it is balanced in the middle of the world." Procopius presupposes that this opinion is true, since he does not argue against it, and he customarily accepts it. Furthermore he clearly insinuates that the ancient philosophers have derived this view from the Sacred Scriptures.

Psalm 103 [104:5] says, "You have established the earth on its foundation (which Aquila and Symmachus read as 'on its seat'); it will not move for all eternity." Commenting on this passage, Theodoretus says, "Since He built the earth on itself, He gave it the property of never moving, and it remains in that condition as long as He will wish. The same is said elsewhere in Job 26 [26:7], 'The earth is suspended in nothingness.'" Thus Theodoretus thought that from the beginning of creation the earth was constructed in such a way that it would remain motionless according to the will of the creator. From this some have rightly inferred that the earth was built on itself and is fixed. That is, it was established in its own natural place, and it is preserved and is at rest where it was born, indeed in the center of the world and in the lowest place in the whole universe. As a result it does not need an external foundation to sustain it, but rather is shaped and solidified by its own heaviness as an internal conserving power, just as the farthest heaven is not enclosed by an external limit but is held together by itself. See the comments in Damascene, *De fide [orthodoxa],* bk. 2, chap. 10.

In "On Psalm 91, 'Qui Habitat,'" sermon 11, St. Bernard says, "The angels are in the highest and a safe place; humans are not in the highest nor in a safe place, but in an insecure place and on solid ground, that is, they are on the earth, the lowest place, but are not in the inferno." If the earth is the lowest place in relation to the heavens, it follows that it is in the middle, and that it plays the role of being an immobile center in relation to other things. When Bernard says that humans have "the lowest place, but are not in the inferno," he uses the common opinion, which locates the place of the damned in the center of the earth.

Finally, lest we go on forever, we quote from the *De confusione lingarum* of Philo Judaeus, who is more ancient than all the other authors referred to above. He says, "Even though all the parts of the earth are either joined together on a small foundation or are raised up in the shape of a column, nevertheless it is separated at the greatest distance from the ethereal sphere, especially according to the opinion of the philosophers who, having studied the matter, strongly assert that the earth is the center of the world." From all this it is quite clear how easily the Fathers would have adopted the old and valid opinion of the philosophers regarding the location and the immobility of the earth. They presupposed it as a certain and proven matter of fact without considering any other view, much less persuading themselves that the sun is located in the center.

Now let us consider the scholastic Fathers. Among others St. Thomas explicitly states that the earth is immobile, round, and spherical in his commentary on *De caelo,* bk. 2, lect. 2. And in his *Opusculum 10,* art. 16, the Angelic Doctor says, "No natural power can move the whole mass of the earth within the space up to the sphere of the moon, for no created power can change the order of the principal parts of the universe, including the fact that the earth is located in the middle. Nevertheless it seems to me that the opposite could be maintained without danger to the faith if the argument were focused on the quantity of the weight of the earth and not on the above mentioned order of the universe." These words should be noted carefully. For St. Thomas thinks implicitly that, if we attend to the order of the universe in which the earth necessarily resides in the middle, then it cannot be said without danger to the faith that the earth is mobile in place, and thus it cannot be placed in a large orbit above the sun. He continues by saying, "But if one considers the local motion, which maintains the above mentioned order, then it seems that the earth is naturally at rest." For Thomas this could be negated only by a power above nature or even by a violent motion contrary to the nature of the earth.

In his commentary on Ecclesiastes 1, St. Bonaventure likewise clearly states that the earth is at rest. Although he speaks also of the duration of the sun's existence, still he says that the sun moves with a special motion. And he thinks of the earth at rest as being contrary to this special motion of the sun, which is circular and twisting, as he explains at length in that passage.

In his comments on Ecclesiastes 1, on Job 9, and on Psalm 92 [93], Dionysius the Carthusian states that the earth remains perpetually fixed in the middle as an immobile center. Even though he is speaking about the duration of the earth, duration can also be attributed to any kind of circular motion. It makes no difference whether that motion occurs at or around a natural center, or if it be any other circular motion that might seem reasonable to some one, even though the views of those who have dreamt of such motions have been specifically rejected.

Finally this view has been held by all those Fathers and ancient authors who read Scripture in the moral sense when it speaks about the earth at rest, for example, Theodoretus, Cassiodorus, Bruno, Regio [of Auxerre], Haimo, Dionysius the Carthusian, and others on Psalm 92 [93]. To these we can add Richard of St. Victor in his *De eruditione hominis,* part 1, bk. 1, chap 1. For all of them deal with the literal sense, on which is based the spiritual sense which includes the moral sense, as St. Thomas correctly points out in [*Summa theologica*], part I, q. 1, art. 10.

CHAPTER 5

It is shown that the same passage of Holy Scripture can have many literal meanings

Although some may deny this claim, it has been overwhelmingly affirmed as true by many others who agree without exception with St. Thomas's most powerful argument in his [*Summa theologica*], part I, q. 1, art. 10, "Since the literal sense is what the author intends, and since the author of Sacred Scripture is God, who grasps all things at once in his intellect, it is not unsuitable that many literal senses are found in the same words of Scripture." This argument is taken from St. Augustine's *Confessions,* bk. 12, chaps. 18, 19, 20, 24, and 31, where the occasion arose for him to discuss at length the literal meaning of the

passage from Genesis 1 [1:1], "In the beginning God created heaven and earth."
Since different authors have given many interpretations to these words, it is
necessary to distinguish the given literal sense from the mystical sense as follows. All the mystical senses must be stated, but one cannot infer which is true
and which is false, or if the mystical sense is sometimes fulfilled, when the
meaning is based on one's free choice rather than on the literal sense. Further
it is necessary to state all the literal meanings, and thus what were the intentions of the Holy Spirit. St. Augustine frequently argues in *Confessions,* bk. 12,
that if all are to agree on a passage, then one cannot determine the mystical
senses, nor any other figure of speech, unless the literal senses are determined
first. No one has been so bold as to assert the opposite of this view, which must
be accepted as true.

This can also be proven from Sacred Scripture itself, for example, from
Exodus 12 [12:47] and Numbers 9 [9:12], "Do not break one of his bones," and
"You will not break one of his bones." This speaks literally of the Pascal lamb,
but is also to be understood of Christ in John 19 [19:31–36]. The latter author
intends to refer to both in his literal meaning, and both are fulfilled, the former
each year and the latter in the passion of Christ, which John expressly states,
"These things were done so that the Scriptures would be fulfilled; do not break
one of his bones." But this could not be fulfilled except in the literal sense.

Furthermore in 1 Paralipomena 22 [1 Chronicles 22:10] the words, "He will
be a son to me, and I will be a father to him," refer literally to Solomon. But
in Hebrews 1 [1:6] Paul attributes these words to Christ, not in an allegorical
but in a strictly literal sense. In general this would not constitute a proof, for
the mystical senses themselves prove nothing, unless they are derived from
the literal senses, as for example, the two wives of Abraham signify the two tes-
taments, and the city of Jerusalem signifies the heavenly free Jerusalem, that
is, the realm of the beatified, eternal life. But it happens on rare occasions that
the one sense transfers to the other.

Further consider Psalm 2 [2:7], "The Lord said to me you are my son, and
today I have fathered you." According to St. Augustine this passage refers to the
eternal generation from the substance of the Father mentioned in Hebrews 1.
But it is also taken in Acts 13 [13:33–34] as referring to the glorious resurrection
of Christ; and in Hebrews 5 [5:5] as referring to Christ receiving the priesthood.
Thus in this case we have not only two but three literal meanings.

In John 11 [11:50] when it is said, "It is better for one man to die for all the
people," Caiphas said this in one sense, but John, instructed by the Holy Spirit,
took it in another sense, as is clear to the reader. The words of Daniel 9 [9:27],

"In the temple there will occur a desolate abomination," are taken by Christ as a prediction of the end of the world in Matthew 24 [24:15–16], but as fulfilled by Antiochus Epiphanes according to 1 Maccabees 1 [1:16–25]. Furthermore the words at Isaiah 53 [53:4], "Indeed he bore our weaknesses," refer to bodily illnesses according to Matthew 8 [8:17], but in 1 Peter 2 [2:24] St. Peter takes them as referring to our sins, a meaning which is also found in the Septuagint.

Finally the words of Habakkuk 2 [2:3] "Up to now he was seen from afar, and he will appear at the end, and will not fail" refer to the first coming of Christ, but in Hebrews 10 [10:38] Paul interprets them as referring to the second coming. Thus to understand many other passages of Sacred Scripture, we should take as true the words of St. Augustine in his *De doctrina christiana*, bk. 3, chap. 27, "In divine discourse what can by inspiration be seen more fully and more completely when the same words are understood in many ways, is approved by divine witness no less than the other meanings."

Further there are as many literal meanings as there are types of equivocation, amphibologies, compositions, and divisions. This happens in the present case, "the earth stands still forever," which can be taken to refer to a state of rest as contrasted to a circular motion, or to perseverence through time. It may happen that the literal sense is permanently fixed and unique because of the word used, and thus it can have only one sense. But if we do not attend to the word used, but to the mind of the author who has wished to signify many things by the same word, then the author has provided many literal meanings, which can be established from other passages of Scripture, and which are moreover fixed and not obscure, as the examples given above show. Otherwise the same word could not be understood to have many meanings. Hence whether the literal meaning be one or many, it is always a definite meaning insofar as it can be taken from the words and in no other way. When there are many literal senses, they are all determined together, and this is known when the Scripture is fulfilled in no way, in only one way but not in another, or in all ways. But when there are several senses that seem to be equally primary, then the literal senses are said to be verified when they are all verified. Or if there cannot be several equally primary senses, as some have thought, then the literal sense is said to be fulfilled when the principal meaning is verified. Thus the quotation given above, "Do not break one of his bones," was verified every year at the Passover, yet it was expected to be fulfilled again. But if this speaks about Christ in its more primary sense, then this principal literal meaning was verified when the event of Christ's death occurred. The same is true of the quotations from Daniel 9 and from Habakkuk 2.

One further point should be noted here. When we said above that multiple literal senses arise because of equivocation, we were not speaking of equivocation proper, which is due to sophistry, and which is properly called a fallacy of language. This occurs specifically when a statement is true in one sense but false in another, thus creating an occasion for error. Rather we were speaking of equivocation in the sense that there are many true and definite meanings, but the intellect either does not know that they are all true, or if reason knows this, it is doubtful which sense really corresponds to the mind of the speaker.

CHAPTER 6

The literal meaning of the proposition "The earth stands still forever"

The full meaning of this proposition consists of many partial truths. One of them is, "The earth stands still." That is, it moves in no way as a whole, or to put it in another way, its does not move from place to place by any sort of imaginable motion, nor change its location or order to things, either in itself or relative to another. This is clear from the words used by Sacred Scripture. That is, the earth is founded, established, made firm, stopped, not moved, established on its own seat or stability, and finally standing and fixed, not moved from its own place or foundation. These words signify that it is completely immobile as a whole. We note also that the proposition, "The earth stands still forever," is rendered in the Hebrew as "The earth is fixed and stopped forever," which indicates a state opposed to local motion. The primary word used here, "established," means "permanent," as when one says that columns are erect and fixed as they firmly hold up a house.

This meaning is the simple meaning, taking the statement in its proper sense and without any ambiguities, so that no appeal to any metaphorical sense is needed. Hence this meaning agrees so much with various scriptural passages that they not only do not conflict with each other, but also they mutually explain each other. This is a sign that this meaning is intended by the Holy Spirit, while on the contrary a meaning cannot be said to be intended if it disagrees with another passage of Scripture. For one truth does not contradict another. See St. Thomas's *De potentia,* q. 4, art. 1. I think this is the reason why the motion of the earth is hardly ever directly mentioned by the Fathers, for there is no room for doubt in this case. Further all the passages of Scripture that seem to affirm the motion of the earth can be easily and only explained as referring to an extraordinary and unnatural motion, for Scripture does not mention any such motion except as a sign of divine omnipotence or indignation or sometimes as a threat. See 2 Kings 22 [2 Samuel 22] and Psalm 17 [18]. The same thing can be concluded from the expressions used in the Hebrew version where the sacred text regularly uses the motions just mentioned in that way, as can be seen in Deuteronomy 28, Proverbs 30, Job 16, John 14, Jeremiah 51, and Joel 2. Sometimes it is also used to express the awe of creatures for the creator, as in Psalm 103 [104], Ecclesiasticus 16, and Isaiah 41. As a result in his *Liber de haeresibus* Philastrius says that those philosophers are heretical and groundless who attribute the motion of the earth to the nature of the elements and of reality, rather than to the special command and indignation of God. In respect to this one should contemplate the teaching of St. Thomas in his *Opusculum 10*, art. 16, to the effect that it cannot be allowed without danger to the faith that the earth moves in its place, if one considers the order of the universe which requires that the earth be located in the middle.

Therefore from this proper sense it follows that the earth cannot move in a large orbit around the sun, nor by a rotation on itself in the same direction as that great orbit, nor by any other contrary motion around its own center, with its axis inclined at any angle whatsoever. For all these cases put the earth in motion as whole, outside of its natural place, and on some axis or other in the universe. The same result follows if it moves accidentally and without its own proper motion by being attached to the orb mentioned above. All these motions openly contradict the literal sense of Scripture in obvious passages from which that meaning is derived.

The proposition, "The earth stands still," has another meaning or partial truth. That is, the earth does not move circularly in its own place around its own center and axis by any kind of motion whatsoever; but it does move in respect to its parts and the differences between them relative to some fixed point, even though the order and arrangement of the universe would not change, nor would the earth as a whole change. This is clear because in general such an inclination toward another direction is contrary to Sacred Scripture: "It will not incline through the ages." Hence for this reason the Latin interpreters have used the words "stands still," which would exclude all motion, and which would encompass the whole literal meaning which is intended here. For the Latin writers understand the words "stand still" to mean a fixation and a denial of any kind of motion, including circular motion, which would be attributed to the earth, whether this refers, as various people have said, to only a cubit of the earth or the largest mountains, or to other unequal bodies, or to a heavy spherical body moved by an easily impeded motion. This is the meaning of the questions, "Whether the earth rotates in a world which stands still?" and "Whether the world goes around the earth which stands still?" in Tullius [Cicero], *Academicae questiones*, bk. 2, and in Seneca, *Questiones naturales*, bk. 7, chap. 2. In general the Latin writers would have easily been able to use another word from the Arabic text, if they had not been considering this kind of immobility, which captures the whole literal sense. Let us add that when the earth is said to stand still and to move comparatively and in the opposite direction relative to the sun, every motion of any kind that is attributed to the earth really belongs to the sun. As a result the earth is devoid of all motions, all of which belong to the sun.

The sun moves primarily in its proper annual motion, but also moves in its opposite daily motion. These meanings are clearly stated in the passage at Ecclesiastes 1 [1:5]. The sun moves and rotates on itself, and this double motion consists of its annual motion around a fixed axis from west to east, and of its almost monthly motion around its oblique axis from east to west. This has been proven by the return motion of sunspots, first discovered by Apelles [Christopher Scheiner], and has considerable support in the same passage mentioned above, "A breath of air travels in a circle and returns in its circle." The latter phrase can easily be understood as a circular motion of the sun on itself. Also the proposition "The sun stands still forever" is said relative not only to the prior statement, "A generation goes and a generation comes," but also to the subsequent statement, "The sun rises and sets," etc. This is clear enough from the emphasis and repetition found in the Hebrew, the Greek, and the old Jerome texts, "The sun rises, and the sun sets," etc. This indicates that the meaning depends on the previously mentioned meaning. Thus when the earth is said to stand still forever, the sun truly rises and sets, and thus the sun as a whole is in motion and is in a continuous motion. All the motions and modes of motion which can and do belong to the sun are opposed to the state of rest of the earth. Finally that this is the literal sense is shown not only from the fact that such a circular motion of the earth has no purpose in nature, but also and especially from the fact that, if any type of circular motion of the earth is granted, then many consequences clearly contrary to the faith follow, as will be shown later.

There is a third meaning. "The earth stands still forever" means that it endures and remains as it is, at least insofar as it is not produced or destroyed like other things. This is explicitly stated in the preceding words, "A genera-tion passes and a generation comes," to which the text adds, "But the earth stands still forever." The meaning here is a comparison, as if it were said that the earth as a whole is neither generated nor corrupted, as are the other things which are generated on it or from it, which refers especially to human genera-tion, as has been explained by Jerome, Olympiodorus, Hugo, Albinus, and oth-ers. Thus what Hugo says about this text is true, "The earth stands still in that it sends forth things that are coming, it carries things that persist, and it re-ceives things that disintegrate." Note also that the Arabic text reads, "Forever it endures or will endure."

I have said that the earth stands still forever insofar at least as it is not produced or destroyed. For as Jerome notes, "forever" or "for ages" can also be taken to mean "for a long but finite duration." Olympiodorus takes this to mean "for a very long time." But in his *Locutiones Exodi,* bk. 2, and in his *First Question on Genesis,* chap. 31, St. Augustine observes that the word "forever" is attributed sometimes to that which has no end, and sometimes to what has an end but we do not know where or when it is. Moreover some have said that it is false to assert and to believe the eternity of the world from this passage of Ecclesiastes, and hence they say that this book is not part of the faith. [Pierre] Galatinus [Colonna] also thinks that this is false in his *De arcanis [catholicae veritatis],* bk. 11, chap. 5, because in the Hebrew text Scripture nowhere speaks of the eternity of the world in the proper sense of the word "eternal." Thus for us it is sufficient to take this word "forever" with the sacred authors as refer-ring to a very long but indeterminate time. So this is how we should under-stand the words: "The earth at least stood still forever," while humans came to inhabit it later. This eternal state of rest is not contradicted by the words of Job 24 and elsewhere to the effect that it is so often changed. For it changes in respect to the image which it now has, as Paul says in 1 Corinthians 7 [7:31], "This world as we know it is passing away," even though it must persist with-out end, and not forming other things but renewing the same things. But it is sufficient for our purposes to agree with Jerome, Augustine, Olympiodorus, and others who have the same or a similar view.

Furthermore, to confirm that these truths constitute the whole literal sense, we turn to the words of St. Augustine in his *De doctrina christiana,* bk. 3, chap. 27. "When from the same words of Scripture not just one thing, but two or more other things are understood, even though it is clear what the author intended, then there is no danger if one is able to show that any one of these meanings agrees with the truth by using other passages of Sacred Scripture.

But he who does this should examine the divine words in such a way as to end up with the intention of the author, through whom the Holy Spirit produced that part of Scripture. And one should either follow that reading, or reject a different reading of those words which does not agree with the true faith, by using the witness of any of the divine passages. Indeed that author perhaps saw that interpretation in the words we wish to understand, and certainly the Holy Spirit, who worked through him, also foresaw without doubt that that interpretation should and would occur to the reader or listener, because he also foresaw that it is based on truth itself. For in the divine words what can be divinely foreseen more often and more widely than that the same words are to be understood in many ways, which are established by other no less divine testimony." So says Augustine.

CHAPTER 7

Which of the above meanings would seem to be the one principally intended

If we focus on the words in the phrase "The earth stands still forever," these words are so specific and definite that they can be taken in only one sense, namely, that the earth is unmoved and unchanged. However, because of the ambiguity and equivocation pointed out above in chapter 5, these words must be understood to refer to a stable condition both in time and in place as opposed to any kind of motion. Therefore all the meanings that we examined in the previous chapter were understood and intended in the mind of the author, as we have shown from the words themselves and from other passages of Scripture. Furthermore of all these meanings some have already been fulfilled and some remain to be fulfilled, insofar as they connote intrinsically or extrinsically some degree of succession. For it is true that the earth endures and does not move in any of the ways listed above. It is also true that it endures forever in the sense of "eternity" as explained above, and it does not move naturally in any way, whatever might have been possible because of divine power, which does not undermine the truth of the literal sense on which the faith is built. For although it is a matter of faith that heaven and earth were created in the beginning, and although they could be returned to nothingness by the power of God, still it is consistent with the literal sense and the truth of faith to deny that this has happened, but not to deny that it will happen in the future. Thus if the power of God has in fact made the earth to endure and to be at rest, then that makes the literal senses true and definite, and it can be affirmed that the earth has endured and has been at rest. And the above mentioned meanings also have been fulfilled, even if later they can no longer be said to be fulfilled because the earth no longer endures and is at rest. Granting all this, it can now be exactly stated what was principally intended among the above meanings, and thus we can answer the question at hand.

I say then that, properly speaking, all of the indicated meanings were equally intended, but accidentally the more principal meaning is that the earth does not move in any way. This is clear firstly for the following reason. Every literal meaning must be true, consistent with the words used, not opposed to tradition or to any other passage of Scripture, and explained by Scripture itself. But as we have seen, all of these conditions are met by the proposition, "The earth stands still forever," insofar as it contains the above mentioned meanings. It necessarily follows that all these meanings were equally and principally intended as constituting one unified truth. Let us add that all these meanings agree when their antecedents and consequences are put together, that they do not depart from the proper use of the language expressing them, and finally that they are fulfilled. This proves that they were all equally intended by the speaker.

Secondly this is clear from many arguments. First, because in many places Sacred Scripture says that the earth endures but in no way moves. Second, because in accordance with our human way of understanding, the meaning of the words "stand still" is more properly taken to signify that the earth has no motion rather than that it persists through time. Third, not one of the Holy Fathers has said that the earth moves, or has doubted that the earth is at rest. Indeed on occasion some of them have discussed this view and the various doubts about it. However they never maintained or proposed that it is true of nature, and they answered doubts, not differing at all among themselves about the motion of the earth. Also they were quite aware of the views of Pythagoras and his followers, and of the opinion of Plato, Aristotle, and other philosophers who agreed with them. A sign of this is that they thought that such views were clearly only dreams, which they held in contempt. Fourth, because the opposing view was not claimed by the Holy Fathers but by others who have no authority in theology, and who are only following their own inventions and imaginations to show themselves off. Putting aside everything that is accidental or merely relative to us or, if you wish, is irrelevant, the meaning that seems to be primarily intended in the Scriptures is, as we have said, that the earth in no way moves, but is at rest in its own natural place devoid of all motion. And it remains fixed in place forever. To this we add a point which seems to prove the matter essentially and absolutely; namely, from the opposite of this view, which opposite claims that the earth moves by itself, many consequences follow which are contrary to truths revealed either explicitly or implicitly in Scripture. We will now prove this.

First, it is a matter of faith that God created the firmament in the middle of the waters to divide the waters from the waters. This firmament has been called the heavens. Then in that firmament He placed two large lights, the larger light (the sun) to preside over the day, and the smaller light (the moon) to preside over the night, in order that they would shine in the firmament of the heavens, and would illuminate the earth by shining upon it, according to Genesis 1. But if the earth were placed and put in motion in a great circle above the sun, Mercury, and Venus (which is the primary meaning intended by the defenders of the motion of the earth), then it follows first that the earth is in the sidereal firmament, and that the sun, which is located in the lowest place at the center, is outside of the heavens and the sidereal firmament, just as the earth is now properly said to be outside of the heavens and the firmament and to be located at the lowest place at the center. Second it follows that the firmament did not divide the waters from the waters. Rather, both the elemental waters, and the denser waters that surround the whole earth in its seas, remained in the firmament, held at almost the middle place. For all those waters either had an inclination to rotate together with the earth in its great orbit, or they remained in the lowest place where they surround and cool the sun. Thus is established the first consequence, namely, that the sun with its surrounding waters is outside the firmament and outside the sidereal heavens, which is clearly contrary to the Scriptures.

Second, it is a matter of faith that the heavens are up and the earth is down. See Deuteronomy 4 [4:39–40], "Know therefore today that God himself is in heaven above and on earth below, and there is no other," and Proverbs 25 [25:2], "In the heavens above and the earth below, the heart of the king is inscrutable," and Psalm 102 [103:11], "His mercy is as large as the height of the heavens above the earth." This way of speaking points to a distance between heaven and earth, with the heavens in the highest place and the earth in the lowest. This is also indicated in Jeremiah 31 [31:37], "If the heavens above could be measured, and the foundations of the earth below examined," and in Isaiah 55 [55:9], "As the heavens are high above the earth, my ways are high above yours." I will omit many other passages, at Mark 13, John 11, Colossians 3, Galatians 4, Deuteronomy 28 and 30, Baruch 3, etc. Also the views on the matter of many of the Fathers are beyond doubt. But if the earth rotates in a large orbit above Venus, Mercury, and the sun, it cannot correctly be said to be "down" in relation to the whole heavens, as is stated in the arguments about the height of the heavens and the lowness of the earth in the passages cited above. For Venus and Mercury would be far below the earth, and the body of the sun at a still further distance, which ought thus to be used as the comparison for the terms "up" and "down." Nor can it be said that this comparison is taken as referring to the relation of the highest heavens to the earth, even though the latter is not in the lowest place. For in the same way one could take this to be a comparison of the stellar heavens to the empyrean heavens, or to any one of the planets which are located far below the stellar heavens. But this is to be found nowhere in the Scriptures, which always refer rather to a simple relation between heaven and earth, as itself being the greatest distance between the highest and the lowest.

From all this I point out in passing how contrary to Sacred Scripture is that system of the universe whose defenders locate the earth in the heavens above Venus and the sun in the lowest place at the center. For from this view it necessarily follows that one cannot accept in its proper sense what is said of Christ in the creed, namely, that he first descended into hell and then ascended into heaven. For what possible figurative sense could be expressed by these words, and what could be more contrary than that to what has been accepted as the common Catholic meaning? Yet the defenders of the Copernican system want to take these words, and many others like them, to refer only to appearances. And thus with this way of speaking they easily overthrow the whole creed in addition to rejecting the truth of these words. And as a result they now invite their audience to revive the Valentinian heresy, which claims that many things did not really happen but are only apparent fantasies. Indeed would not the eternal Truth make a mockery of us if figures of speech were used in the established articles of faith and in what we need to know for salvation, in such a way that we do not know what we believe, and even that in reality the opposite is what happened, while we persuade ourselves to believe otherwise? Furthermore if these words and others like them are spoken in terms of what is apparent to the senses, then the opportunity will be presented to use right reason to grasp the truth as it is, to form true propositions accordingly, and with the aid of interpretation to reduce Sacred Scripture to its true and proper sense.

Thus what is said in John 11 [11:41], "Jesus with eyes raised upwards," would be expressed as, "Jesus with eyes bent or turned downwards." And Acts 2 [2:19], "I will show marvels downwards in the heavens, and signs upwards on the earth." And Colossians 3 [3:1–2], "Seek things that are below, where Christ sits at the right hand of the Father; know the things that are below and not above the earth, that is, know the things which are upwards." Finally what Christ said to the Scribes and Pharisees at John 8 [8:23], "You are from below, I am from above; you are of this world, I am not of this world," would be converted to this truth, "You are from above, I am from below; you are of this world (taking "world" according to the Ptolemaic system); I am not of this world, but of the world according to the Pythagorean system." Away with these trifles and heresies. Nevertheless they follow from the opposing opinion, whose defenders maintain that Pythagoras derived his teaching from Moses. Is it not incredible that these defenders are not persuaded that the Scribes and Pharisees often objected to Christ and to Moses, unless they take the above mentioned passage to mean that Moses, Pythagoras, Christ, the Scribes, and the Pharisees all completely agreed about the system of the world? But let us dismiss these absurdities.

Third, it is a matter of faith that God suspended the earth above the void, as is said in Job 26 [26:7]. The Fathers, Ambrose, Isidore, Damascene, and others deny that humans can understand whatever happened. This is also a matter of faith because of what is said at Job 38 [38:5], "Do you know who can measure it?" and at Jeremiah 31 [31:37], "If the heavens above could be measured, and if the foundations of the earth below could be investigated." If the earth were located in a large orbit, and therefore in the sidereal heavens, it could not be established above a void, just as the other stars and stellar bodies located in the heavens are not established above a void, as is easy to understand. Indeed no body of any kind located in the heavens can be said to be suspended above a void. And it is not difficult to investigate and measure all these bodies which can be seen, whether that other heaven be a solid and impenetrable body, or (as seems more certain to me) a fluid composed of air or water, that are the various opinions that have been maintained on this topic.

Fourth, it is a matter of faith that during the suffering of Christ the sun experienced a miraculous disappearance, and for three hours, that is, from the sixth to the ninth hour (Matthew 27 [27:45]), darkness descended on the whole earth. But if we assume the motion of the earth which its defenders imagine, this could not have happened in a rational way.

For since in its daily rotation the earth would have moved forty-five degrees towards the east, it would be necessary to exempt the city of Jerusalem from that motion. Otherwise one would have to say that the whole of Judea was in darkness, while the moon remained in its place at rest under the sun. Likewise all the regions to the west would have successively lost their light in such a way that darkness did not descend all at once upon the earth, but only on successive places over three hours. But this makes a mockery of the meaning of Scripture and of the views of all its interpreters. And if it is claimed that the moon continually moved backwards with the earth toward the east for forty-five degrees, then that motion is contrary to nature. For it would immediately follow either that another miracle occurred that neither Scripture nor its authors record; namely, that the rays of the sun were prevented from illuminating, or that the eclipse was not seen in the west because, while the moon was in an intermediary position, the sun was uncovered more quickly as the moon moved towards the east.

It also necessarily follows that darkness did not descend on the whole globe of the earth, which is the common opinion of all interpreters contrary to Origen and a few others, and which is proven by the oracle of the Sibyl, by the words of Phlegon, by the Greek commentators, by the witness of the widely dispersed Echnicans, and finally by the best judgment of the mathematicians. This is especially true since many miracles would be needed for the moon to block the sun, because the moon would move backwards toward the east with its own oblique motion and path, rather than with the different oblique paths and angles needed for it to intersect with our line of vision. And as a result it would block a smaller part of the sun than would be required to plunge the whole globe of the earth into darkness. For when the moon is in the way of the sun and eclipses its whole disc, the diameter of its shadow is no larger than 350 Italian miles, while in the present case the darkness would cover much less space. All these arguments are based simply on the theory of plane triangles.

In addition to this is the argument from the principles of optics that the sun illuminates more than half of the surface of the earth. Hence if the earth moves, and the moon regresses, and the sun stands still, then it would hardly be possible for the darkness to last for three hours without a new miracle, as we said. This is so clear to anyone who considers the matter that no other argument is needed here; otherwise this will become a mathematical treatise. Moreover we think that this argument is very powerful for whoever believes that the darkness is truly caused by the interference of the moon and that it is a mistake to deny that the moon moves in its own path. There are indeed some who think otherwise, that the sun fails to shed light on the earth because the path of the moon is divinely decreased to conform with the sun.

Fifth, it is a matter of faith that the sun moves, and that it moves in a circle, as we will prove in chapter 9. Moreover it moves around the earth, as is clear both from the physical and the mathematical facts and from passages found in Scripture. It necessarily follows that the earth is located in the middle and is devoid of all motion. Otherwise, if it were in motion and if nothing else is immobile, then what motion is the final motion in the present order of the universe, for that cannot be attributed to the motion of the sun? That the earth is at rest is not only a direct matter of faith, but it is also a matter of faith insofar as it is immediately deduced from another proposition of faith, namely, that the sun moves in a circle, in which the earth is properly and virtually contained. Perhaps the former is more certain, since it is also a matter of faith that the earth is the center of the universe, which must be granted, as we will see below in its proper place. This is also the principally intended literal meaning of the stability of the earth, according to our mode of understanding, as was explained above. Furthermore things that are opposed to the motion of the earth are easily deduced from the Scriptures, and many of them are obvious and completely clear to the senses. From this it is further deduced that, since the motion of the sun is a matter of faith, this destroys the system of those who claim that the sun is at rest at the center, and that it is devoid of any rotation on itself, which the authors of that view attribute to the earth.

CHAPTER   8

The unanimous opinion of all the Holy Fathers and Interpreters is that the sun moves

The absolute motion of the sun is asserted by all of the Fathers and ancient interpreters, to say nothing about the more recent ones. They all understand the miracle of Ahaz's clock, mentioned in the passages of Scripture at 4 Kings 20 [2 Kings 20:9–11], Isaiah 38 [38:8], Ecclesiasticus 48 [48:23], and 2 Paralipomena 32 [2 Chronicles 32:25], to refer to the sun's true motion and true retrogression. See Origen, *Homily 3 on Matthew*; Gregory Nazianzenus, *Oration 19*; Augustine, *De civitate Dei*, bk. 1, chap. 8 and bk. 1, chap. 28, on the author of the wonderful Scriptures; Jerome on 2 Paralipomena 32 [2 Chronicles 32], and on Isaiah 7 and 38; Cyril of Alexandria on chapter 39; Theodoret on Psalm 20 [19]; Procopius on Isaiah 38; Glycas's *Annals from Hippolytus and Eustathius of Antioch*, part 2, page 271; Angelomus's *Stromata* on 4 Kings [2 Kings]; Dionysius the Carthusian on Isaiah 39, art. 67; Bede, Lyranus, Hugo, and Abulensis on 4 Kings 21 [2 Kings 21]; and Thoringus, the reputed author of the *Commentary on the Glossa Ordinaria*, who is also well known as the Burgundian. It would be too tedious to quote the words of each of these writers.

Also of relevance are other passages of Scripture that the interpreters understand as speaking of the motion of the sun. See Dionysius the Areopagite's *Letter to Polycarp*, and Justin's *Refutation of Aristotle's De caelo II*, and his response to question 59 [in *Ad orthodoxos*] where he says the same thing. And in his comments on Psalm 134 [135] Chrysostom says, "The sun crosses through the heavens in a whole day, but a lightning bolt goes across the whole orb of the earth in the smallest moment of time." See Rabi Hacados's comments on Genesis 1. In his letter 100 to Paul Isidore of Pelusium speaks of the path of the sun, even though he says that the study of this and other topics in astronomy will produce little of value. And Basil also speaks of the path of the sun in his *Homilies on the Hexameron*, 2, 6, and 9, and in his commentary on Isaiah 13. So does Ambrose in his *Hexameron*, bk. 4, chaps. 5 and 6; and Augustine in his *Homily on St. Vincent Martyr*, and in his *De Genesi ad litteram*, bk. 1, chap. 19. In his commentary on Genesis 1 Procopius maintains that the sun is moved and is not located in the first heaven. And on the words, "God called the light day," he says, "The light completed the path which the sun now crosses." On Ecclesiastes 1 Bonaventure teaches that the sun moves in a special way, while the earth stands still. See also Dionysius the Carthusian, *Opusculum de consideratione theologica*, art. 57, and on Ecclestiates 1. And on Isaiah 39:12 Theodoret says in his *Sermo 3 de angelis* that the sun is carried from its rising and completes the day. And he maintains the same thing in his *Sermo 5 de materia et mundo* against the Greek infidels, when he says, "The sun cannot melt or dissolve the most firm heaven, through which it continuously runs; nor does the firmament itself, even though it is water by its nature, extinguish the blazing sun." And it must be noted, as pointed out by Dionysius the Areopagite in his *De divinis nominibus*, bk. 1, chap. 4, and in his *Letter to Polycarp*, that the path of the sun is maintained by the principles of nature in such a way that it can be broken only by a higher force. The same point is made by Dionysius the Carthusian in his *Opusculum de consideratione theologica*, art. 57. And long before this in his *Origins*, bk. 3, chap. 49, Isidore of Seville said, "The sun moves by itself and is not turned by the world, for if it were fixed in the heavens, every day and every night would be equal. But we see that the sun will rise in a different place tomorrow, and that it set in a different place yesterday. Thus it is clear that it moves by itself and is not turned with the world." So says Isidore.

In his comments on Psalm 90 Chrysostom says, "The sun standing in the center is called the middle of the day." He is speaking of the sun at its vertex or highest point, where it seems as though it stands still because its motion is the least perceptible to our senses because of its very great height. The astronomers call this height, among other things, the center of the hemisphere, where the middle of the day occurs when the sun is there. Thus it is said in Joshua 10 [10:12] that, "The sun stood still in the middle of the heavens," that is, under the meridian or the highest point in the middle of the heavens, as Sedulius has explained. According to Augustine the author of the wonderful Scriptures says that, "Joshua petitioned the Lord that the sun would not move when it was located in the middle of the day." Or perhaps "the middle of the heavens" should be interpreted in Tertullian's sense in his *Ad Scapulam*, where he says that the sun is located in its own high place, and thus could not undergo a prodigious loss of its light because of some ordinary eclipse. We will concede this in the light of copious chronological demonstrations. At any rate Chrysostom did not think that the sun actually stands still, for such a true absence of motion would otherwise contradict his words which we have quoted just above from his comments on Psalm 134 [135]. In the same way Augustine said in his *Homily on St. Vincent Martyr* that the sun seems to really stand still, but nevertheless it moves with the greatest speed. He says, "It seems to you that the sun does not move, but still it moves. Perhaps you will say that it moves, but slowly. What reason do you offer why we do not see that?" He then continues on to explain the speed of its motion. Thus also Nazianzenus said in *Oration* 34 that to reason the sun is stable, devoid of motion, and also indefatigable. He asks, "What drives it on and rotates it with perpetual motion, while to reason it is stable and devoid of motion, is truly indefatigable, the giver of life, and the father of living things," etc. This mode of speaking presupposes motion, but not variation. This can be understood to apply to regular motion, which always occurs uniformly, although it sometimes appears to us to be irregular. And for that reason what is in constant and invariable perpetual motion can be said to be devoid of motion. For in that apparent inequality of motion there is, as Nazianzenus says, "a marvelous equality in which is preserved the divinity, which, although always in action, consists of the eternal and tranquil rest of beatitude."

These are the views of the Fathers who speak of the motion of the sun in an unqualified sense. There are also very many Fathers who could be mentioned who have discussed its circular motion, although in different senses. Some of them deal with the passage of Ecclesiastes 1 [1:5–6], "The sun rises and sets, and returns to its place where, reborn, it rotates through the meridian and curves to the north," etc., which they understand to refer to the daily motion of the sun; others discuss its annual rotation; and finally others speak of unqualified circular motion. The daily motion is discussed by Gregory of Nyssa, Gregory Neocaesar, Augustine the Chaldean, St. Thomas in his comments on Job 37, and Chrysostom in his *Homily 12 to the People,* where he observes that the sun and the other stars revolve daily in a motionless heaven. The annual motion is explained by Jerome, Olympiodorus, Theophilus of Alexandria's *Epistula 3 Paschalis, Glossa [ordinaria],* Hugo of St. Caro, Hugo of St. Victor, Albinus, Bonaventure, and other later interpreters, who maintain that the Hebrew word which explicitly means "to rotate" is expressed by the word *flectitur* ["is curved"]. This indicates an oblique motion between the tropics which is not inconsistent with the daily motion. Finally unqualified circular motion seems to be discussed by Dionysius the Areopagite in his *Letter to Polycarp,* Junilius [Africanus] in his *Commentary on Genesis,* Rabi Hacados on the same text, and some others. But among all the others Claudius Mamertus has elegantly portrayed all these motions in his *De statu animae,* bk. 2, chap. 13. He says, "In regard to the visible heavens it is forever beyond us humans to understand the heat and the light of the sun, the changing increases and decreases of the lunar globe, the fixed and the wandering circuits of the stars, the motions of the stars through their great circles, or how the fixed sequence of day and night varies, or why an alternating influence of heat and coldness warms the world, or what fixed dimensions mark off the parts of time, or how the paths or the lines of these motions return to the same place and are brought back into the same circular paths, or how the ether is decorated with its distinctive numbers and times and melodies." So says Mamertus.

Therefore according to the opinion of all the Fathers it is certain that the sun moves in some way or other, and thus it cannot be at rest at the center of the universe, even though one were to hold that some of the planets move around the sun in their own particular circles. This argument is enough to establish the truth on this issue, even if we were to have nothing else from Sacred Scripture. I also reject those who have maintained that the heavens indeed move circularly, but who go on to add wrongly that all the bodies in the heavens return to the same place in 36,000 years, and that this causes the return of the same effects in things which exist now. This error has merited condemnation in Paris. See Anathema, chap. 5, num. 2, against various errors of faith, by Etienne Tempier, Bishop of Paris.

CHAPTER 9

The sun moves, and its circular motion is a matter of faith [de fide]

That the sun unquestionably moves is established by Joshua 10 [10:12–14], "'Sun, do not move over Gibeon' . . . and the sun and the moon stood still in the middle of the heavens. Never before or after has there ever been such a long day, when God obeyed the words of a man." These words presuppose that the sun moves, and that it has done so since the beginning of creation when it was ordained to reside over the day as the greater light. This is why the above text says that "never before or after has there ever been such a long day." The sun's motion follows from the fact that it miraculously stopped in its motion, precisely when God obeyed the words of a man. And in Ecclesiasticus 43 [43:2–3], "The rising sun proclaims its view of the admirable container made by the Most High; at midday it burns the earth, and who can bear to look at its flames." This text clearly says that the sun rises, and that it moves to the middle of the heavens. And a little later [43:5] it is said, "Great is the God who made him and who maintains its path by his words," which was acquired at creation. And in Isaiah 38 [38:8], "And the sun moved backwards through ten spaces." The same thing is said in 4 Kings 20 [2 Kings 20:8], "So the prophet Isaiah called upon God, who drew back the shadow through a distance of ten spaces through which it had already crossed."

I realize that some prefer to say that this miracle happened only to the shadow and not to the body of the sun, for the sun is not mentioned in our text of 4 Kings 20 [2 Kings 20], nor in the Hebrew text. The same is true of Isaiah 38 [38:8], where the same story is repeated in a briefer version, "Behold, I will move the shadow backwards through ten of the spaces which it has crossed on Ahaz's sundial." But these passages are to be interpreted in the light of Ecclesiasticus 48 [48:26], "In his [Hezekiah's] day the sun moved backward and prolonged the life of the king," and also 2 Paralipomena 34 [2 Chronicles 32:31], "Messengers were sent from Babylon to inquire about the extraordinary event which happened above the earth." The Babylonians clearly did not think that the miracle had affected only the shadow of Ahaz's sundial. Finally in Isaiah 38 [38:8], immediately after the words just mentioned, which I believe gave rise to this dispute, the text adds the words that we first quoted, "And the sun moved backwards through ten spaces which it had crossed." And although 4 Kings 20 [2 Kings 20] does not mention the sun, that makes no difference. For it is justified to infer the cause from its effect, especially since throughout his long life Hezekiah claimed that this sign was that the shadow moved backwards ten spaces, and not that it increased by ten spaces. Therefore this effect would not have happened without the operation of a cause, which was the retrogression of the sun. The same thing is repeated at Isaiah 38 [38:8]. When Isaiah responds to Hezekiah's wish and request, he speaks literally, as I will explain, where the text says, "Behold, I will move the shadow backwards through ten of the spaces," which it had crossed in Ahaz's sundial. But then it is immediately added, "And the sun moved backwards through ten of the spaces which it had crossed," which implies the operation of the cause which produced the miracle.

The fact that the Holy Fathers, and others I could mention, have interpreted this passage as referring to the motion and the retrogression of the sun is so obvious as to need no proof. That is why St. Caspar rejected another interpretation that he judged to be rather improbable, deciding that one should not turn to another opinion when the authority of almost all the Fathers and exegetes is opposed, no matter how ingenious it might be.

Next the fact that the sun moves through a circular path is established by Ecclesiastes 1 [1:5–7], "The sun rises, and sets, and returns to its place; reborn there, it rotates through the meridian. And the wind is reflected to the north, carrying all things in its path, as it rotates in circles." This passage has been taken again and again to mean that the sun moves in a circle, but that has been understood to be either the daily or the annual motion, according to various interpreters and the uses of other passages in Scripture. This same difficulty is also found in the Hebrew, the Greek, and the old Latin text of Jerome. "The sun rises, and the sun sets" indicates a continuous motion without interruption, such that the whole is said to go and to return. This is clearer if with the Hebrew text we read this in the future tense, "The sun will rise, and the sun will set." This properly indicates both a tireless continuous motion and the attribution of that motion to the sun and not to the earth. That is emphasized by the use of the word "sun," which is repeated. At the end of the text the sun is referred to by the word "wind" [*spiritus*]. This is either because of the fiery nature of the sun, or because of the intelligence which helps it, or because of the speed of its motion, which Olympiodorus says is the literal meaning. Or finally this could be because of its effect, that is, the life which it breathes into things, since as it completes its annual cycle, it makes things grow and breathe and flourish. But this is not because the sun itself has an informing and living soul, as was thought by the pagans according to Olympiodorus, and by Origen of all celestial bodies in his *Periarchon, On First Principles*, bk. 2, chap. 7, "On the Fiery Nature of the Sun." See the writings of Eustathius of Antioch, who has been revealed as comparable to the familiar learned scholars because of the elegant Greek to Latin translations recently made for the first time by Leo Alacci. For the same reason the ancient writers designated the sign of Leo as the stopping place of the sun, taking the lion's courage to signify the sun's power of action, which without doubt arises from its hot and fiery nature, as we have said. Eustathius argues that the sun's nature is produced by the union of light and fire, for the light which illuminated the heavens before the formation of the sun is later collected together in the body of the sun, where it governs by burning and illuminating. Almost the same view has been derived from Zeno by Alexander of Lycopolis, a most outstanding philosopher, in his *Adversus Manichaeos*, which Leo Alacci has also recently translated into Latin. Those who think that the stars are composed of the elements of fire, air, and water would not be strongly opposed to this view. See the commentary on the *De causa causarum* by Isaac of Seleucia, which I have seen but which has not yet been translated into Latin from Chaldean. In oration 5, chap. 7, where he presents this opinion about the stars, he eliminates the element of earth from their composition, as if God had said, "To make the stars less dense, the lowly matter of earth will be excluded from them." But there is no need to examine here all these things which are contained in this passage. The main point is that the Holy Fathers agree on the view that there is nothing in common between the earth and the stars and planets, and that the earth by its nature was placed in the middle of the universe, and therefore in the lowest place.

Secondly the same thing can be proven from 3 Esdras 4 [1 Esdras 4:34], "Great is the earth, and high is the heaven, and swift is the course of the sun, which rotates through its path in a circle across the heavens each day." This book, along with 4 Esdras, was excluded from the list of canonical books authorized by the decree of the Council of Trent. Nevertheless it has very great authority, and both books are cited by the Holy Fathers frequently and as above other noncanonical writings. Indeed they are found in some Latin Bibles, both printed and hand-written, immediately after the other books of Esdras. In many Bibles today they are separated off and included in the same volume at the end of the canonical books. Nevertheless for our topic this passage ought to have great authority because, as Olympiodorus points out, this opinion is derived from Ecclesiasticus 1, already quoted above, which states that the sun moves in a circle, which is the interpretation of others also. In chapter 13 below we will present and resolve the objections which some have raised against this passage, along with other arguments.

CHAPTER 10

Whether it is a matter of faith [*de fide*] that the earth is the center of the universe

If it is a matter of faith that the heavens, which contain all things, have a circular and spherical shape, then one could perhaps easily infer that it is also a matter of faith that the earth is the center of the universe, especially in the light of those passages of Scripture which we used in chapter 7 to show that it is a matter of faith that the earth is located at the greatest distance downwards from the height of the heavens. For it is necessary that there be some center of the celestial circle, since a center is by definition the center of a circle. Given that distance, then that center is the earth, which thus was placed in the middle as equidistant from all the parts of the heavens, in relation to which the earth is said to be down and the heavens up.

Although it has been satisfactorily established by mathematical arguments that the heavens have a spherical shape, as we have said in chapter 2, nevertheless this was not accepted by the Fathers, who thought differently because of various passages of Scripture. Many were not persuaded but held this view only provisionally. Others thought that only the planets, for example, the sun and the moon, seem to move around the earth, and not the whole heavens. Finally others thought that while the heavens stand still, the sun and stars move in circles, either as a matter of fact or as a view which can save the appearances. This was favored by Chrysostom in his *Homily 12 to the People,* and by

Augustine in his *De Genesi ad litteram,* bk. 2, chap. 10. Therefore for these reasons it does not seem certain that the earth is the center of the universe, as would be required if this is a matter of faith.

Although it is indeed a matter of faith that the sun moves in a circle, nevertheless the commonly accepted view is that it has an eccentric orbit. As a result the earth cannot be properly said to be fixed at its center either as a whole or indeed in respect to any part of it, since that eccentricity extends over 38 earth radii according to Al Battani, or 48, approximately one-quarter more, according to Ptolemy.

I do not think that these facts are any obstacle to saying that the spherical shape of the heavens is rather probably a matter of faith. First we have already shown that that clearly follows from the circular motion of the sun, which is a matter of faith. For it is completely appropriate that the motion of the sun would be perfectly symmetrical with the shape of the heavens. Nor is this view overthrown by the objection mentioned above, that the orbit of the sun is an eccentric, while the earth has the nature of a center, not in regard to part of itself, but as a whole which is concentric with the whole world. And thus the true motion of the sun is always referred to the center of the universe, from where we observe it, even though its average and apparent motion is primarily related to the center of the eccentric, on which falls the angle of the true motion above the center of the world.

Second our thesis follows from Ecclesiasticus 24 [24:5], "I alone have rotated through the circle of the heavens and have penetrated the depths of the abyss." There is no reason not to take these words as speaking about the heavens simply and properly and without any figure of speech. And in chapter 43 [43:11–12], "See the rainbow, and bless him who made it. Exceedingly beautiful in its splendor, it crosses the heavens in the circle of its glory, with the hands of the Most High exposing it." In the same way these words should be taken as speaking literally about the rainbow's similarity to the heavens in its path and its circular shape. Hence one must infer correctly from these words that the heavens are round and spherical.

St. Jerome agrees with this when he says, while commenting on the words, "what is the breadth and length," etc., in Ephesians 3 [3:18], that it can be concluded from Ecclesiastes and many other passages that the heavens are round and are rolled up in the shape of a sphere. He does not identify the passage, but seems to refer to chapter 1 [1:5], "The sun rises, and the sun sets, and returns to its place." For it is proper for a circular motion to return to the same place from which it started; furthermore a circular motion is proper to a spherical body, other things being equal. Jerome says he used many passages to argue from the circular motion of the sun to the roundness of the heavens. Likewise from this same circular motion as harmonious with the figure of the whole, we have deduced that the heavens as a whole are round and spherical. We have explained above in Chapter 2 why some of the Fathers seem to have thought differently. Certainly St. Augustine, who was able to lead an army for the contrary view, changed his mind later to the truer opinion, as is clear from what we have said in chapter 2, and as is clearly found in his *De Genesi ad litteram,* bk. 1, chap. 10.

All the Fathers who think that the heavens as a whole move are in agreement with us. Among them Dionysius the Areopagite is the highest authority after the canonical Scriptures. In his *Letter to Polycarp* he expressly states that the heavens move, along with its contents. He gives two accounts of the fact that the sun stopped in the middle of the heavens at Joshua's command. Either the entire machinery of the celestial globes stopped along with the spheres of the sun and the moon, or only the sun and the moon stopped while the remaining spheres, both above and below, continued on their usual paths. This second account seemed better to Dionysius; but the first account seemed more likely for others, since it did not change any of the relationship of the spheres to each other. For our purposes each account is likely, and each is affirmed by Dionysius. The other Dionysius [the Carthusian] speaks of the whole heavens when he says that the circular motion of the heavens is natural in his *De consideratione theologica,* art. 57, and he proves this from the earlier Dionysius' *De divinis nominibus,* bk. 4.

There are other Holy Fathers and some scholastics who either assert or prove with arguments that the heavens are round and move in a circle, except for the empyrean heaven which has no motion. St. Augustine does not disagree with this, even though he seems to say in his *De Genesi ad litteram,* bk. 2, chap. 10, that if the heavens did not move, the stars could still move and perform their duties. For in answering the objection of those who ask how the heavens can move when they are called the "firmament," he replies, "The word 'firmament' does not mean that we should think that the heavens stand still; for 'firmament' should be taken to mean not 'stationary,' but either 'stability' or that the heavens are an impassable barrier between the higher and the lower waters." This argument has also been used by philosophers and mathematicians who have confirmed this view with various proofs.

Finally this view is accepted by all those Fathers who think that the earth is located in the center of the universe, as we have discussed above in chapter 4.

In his *Hexameron,* homily 1, Basil says, "Our discussion has shown that the middle of the universe is the lowest place." He is speaking here specifically about the earth, as is clear to anyone who reads the text. He says the same thing in homily 9. In his commentary on Psalm 118 [119], octuple 12, Ambrose concludes that this same view is found in the Sacred Scriptures. He says that this is not only his own opinion, but also that of the pagan philosophers, and more importantly it agrees with Sacred Scripture. Anastasius of Sinai also agrees. In his *Anagogica contemplatio in Hexameron,* bk. 1, he makes many claims, including that the earth is the center of the universe, as he learnedly refutes the view that, "The earth is the source of the universe, since it is the center of the celestial sphere, and hence is older than that sphere or circle." These words, he says, are to be taken logically and not physically, as though God had not created the heavens and the earth at the same time, and had not worked supernaturally but according to the natural order and human powers; and as though it had not been said that "God created heaven and earth," but rather, as he says, "Since the earth is the center of the sphere, one indeed must say not that God created heaven and earth, but that he created the center and then the sphere."

Claudius Mamertus also agrees. In his *De statu animae,* bk. 2, chap. 4, he says, "The earth is certainly the lowest of all the elements; for clearly in a sphere the center is the lowest point, because that point is furthest from the sphere." Finally St. Thomas also agrees in his *Opusculum 10,* art. 16, as we have seen in chapter 4. In dealing with the structure of the universe, he says that the opinion that the earth is located at the center cannot be denied without danger to the faith, since it is said that the whole mass of the earth could be moved from its place by an angel.

Let us grant the following. Since the circular motion of the sun is a matter of faith, then it is rather probable to infer from this that it is also a matter of faith that the whole heavens move with the same motion, and hence that the heavens have a spherical shape. Furthermore it is a matter of faith that the earth is located in the lowest place, and it is certainly stated in Scripture that the heavens move around the earth. But circular motion requires some immobile point around which it moves. As a matter of fact this immobile point is the earth, which also as a matter of faith is in the lowest place. Thus according to the opinions of the Fathers, it is rather probable that it is a matter of faith that the earth is located in the middle of the universe, and thus it plays the role of a fixed point. Granting all this, as I said, it is rather probable that it is a matter of faith that the earth is the center of the universe, which is sufficient at this point for our present purposes. I say that this is "rather probable" because, if one were to construct from what has been said any kind of a syllogism in which both premises were certain as a matter of faith, then its conclusion without doubt will also be certain as a matter of faith. Furthermore when I say that the whole heavens move in a circle, I do not deny its fluidity. For both of these can be true at the same time, as if the earth were surrounded by a fluid sea. As the sea is enclosed within its limits, so are the heavens within its limits. This does not restrict its rotation, but prevents it from flowing downwards. If according to the mode of speech used by the Fathers there are many distinct spheres, there are no separations between these solid spheres, and they describe the periodic paths of the starry heavens and of the stars and planets in the fluid.

CHAPTER 11

In what way various propositions have been accepted up to now as matters of faith [de fide]

First Proposition. "The earth stands still forever," in the sense that it has no local motion as a whole, is a matter of faith, not directly or primarily, but indirectly and secondarily, yet still immediately. The first point is clear. That the earth stands still forever in the way stated is not primarily proposed to be believed as a fundamental claim of the Christian faith. The second point is clear. Since something contrary to the faith would follow from the opposite of the given proposition, that opposite would falsify the passages of Scripture from which the given proposition and others connected to it are deduced. The third point is clear. For the divine witness is open and simple and of the same type in all propositions, even though we perhaps have not adequately taken note of it, which is accidental. Furthermore this proposition is immediate because it is not contained in any other revealed proposition, either as a part in a whole, or as an effect in a cause, or as formally contained in exactly the same claim, or in any other way.

Second Proposition. "The earth stands still forever," in the sense that it has no circular motion, as stated by its defenders, is a matter of faith, not directly and primarily, but indirectly and secondarily. The same argument which applied to the first proposition also applies here. The opposite of the second

proposition implies that the sun is at rest and that the earth is not firmly suspended in nothingness. These views are contrary to the faith. This proposition
is a partially immediate matter of faith, insofar as it is contained in those passages of Sacred Scripture that deny that the earth moves locally as a whole because of its stability, even though some defenders of the second proposition
attribute such a motion to the earth as circulating through the poles of the
heavens in a great circle.

Third Proposition. "The earth stands still forever," in the sense that this refers
to its duration, is an indirect and secondary matter of faith for the same reasons.
For its opposite implies that the Scripture uses a false analogy when it says that
generations go and come. For that reason this third proposition can be said to
be a partially immediate matter of faith, insofar as it is clearly signified as the
intention of the speaker. And it is partially mediate, insofar as in the proposition "Generations go and generations come" there is contained, either virtually
or as arising from our mode of thinking, a disposition to infer this third proposition with necessity from its conceptual implications.

Fourth Proposition. "The heavens are up, and the earth is down," is a matter
of faith which is direct, primary, and partially mediate. The first point is clear.
Since this claim is contained in the propositions that signify that Christ descended to earth to procure human salvation and, having done that, he ascended back into heaven, which is an article of faith, this fourth proposition
must be believed under these terms in order to obtain salvation. And if it has
been denied, one has destroyed not only many things contained in Scripture
in various ways, but also the articles of faith, which are proposed for belief as
necessary for salvation, either explicitly or implicitly. The second point is clear.
The facts that the heavens are up and the earth is down are completely clear
and obvious, and thus are presented with sufficient attention, which is what it
means to be revealed directly and immediately. Furthermore this is contained in
the propositions that constitute the articles of faith, as either completely iden-
tical with one of them or as a part in a potential whole. For indeed if Christ ascended into heaven, he also ascended upwards, as in Colossians 3 [3:1], "Look
toward the things which are above, where Christ is at the right hand of the Father, etc." This is what it means to be revealed mediately.

Fifth Proposition. "The sun moves," either absolutely or by rotation, is a matter of faith indirectly and secondarily, but immediately. The first point is clear.

This is not an article of faith, but its opposite implies something that is contrary to faith. For since there are many passages which expressly say that the sun moves, to deny this is to maintain that the Scriptures are false. Furthermore in many ways this would create a danger of misunderstanding God's providence. Nor could it be absolutely said in Romans 1 [1:20], "Through things which have been made, the mind can see divine things which are invisible to the created world, as well as his eternal power and divinity." These words are verified by any creature, and especially by a consideration of the sun, which is clear not only from various passages of Sacred Scripture in Genesis 1 and Ecclesiasticus 43, but also from the Holy Fathers and various ancient philosophers whose opinions have been guided by the light of nature. The second point is clear. First this proposition is evident from many clear passages of Sacred Scripture which need no explanation, by the subject matter itself and the end for which the sun was created, and by the evident fact that it was established with such qualities. Finally the sun's motion was revealed in itself, and not virtually or formally or in any other way contained in another primary and immediately revealed proposition, which would mean that it must be said to be mediately revealed.

Sixth Proposition. "The earth is the center of the world" is rather probably a matter of faith, as we have said, and it is indirect and mediate. The first point is clear. For the opposite implies that the earth is not suspended in nothingness, and that it is not at the most remote distance from the heavens (contrary to many passages of Scripture), while a center must be at the maximum distance from the surface of a sphere or from the circumference of a circle. Moreover some other immobile body would then have to be located in the center of the universe; but this is hardly possible according to the present order of things, leaving out the judgment of the faith. This is especially true if the body of the sun were located in the center, or if it be some other body which now moves above or around or adjacent to the earth, or which moves in any other way. The second point is clear. That the earth is the center of the universe is contained in those propositions from Scripture, in Ecclesiasticus 24 and 43, which imply that the heavens have a spherical shape, which is proper for the most perfect body and for its motion, as we explained in the previous chapter.

Furthermore we wish to advise the reader that in this discourse up to this point, we have used syncategorematic terms, such as "mediately," "immediately," "directly," "indirectly," etc., according to accepted theological practice when we have used them to explain our opinions. We say this so that no one can by chance take them differently according to other usages, and thus fall into an interpretation different from what we intend and less accommodated to the subject matter.

CHAPTER 12

Given that the earth at rest is a matter of faith [*de fide*] whether it is allowable to argue for the contrary

First I say that, if it is a matter of faith that the earth is at rest, and if this has been adequately promulgated, then it is in no way allowable to argue for the contrary. This is clear because, if it is not allowable to doubt this, then much less can one impudently dispute it because of the danger thus created. Otherwise indeed the agitation and heat of intense debate would easily cause a lapse into manifest and obstinate errors, resulting in not only private dangers of bad thinking but also in the public danger of enticing others to reject the opinion. It is difficult enough to hold onto the established opinion that one accepts in every way, and as soon as you reject it, you will receive a reward from those who have too much time on their hands. As I will soon explain, these dangers have a tendency to arise much more from the present topic than from the other topics which are regularly disputed in philosophy and theology, whose truths are so religiously agreed to by all learned men that no Christian would ever doubt them.

Second I say that it is allowable to assume the motion of the earth hypothetically for the purpose of making mathematical calculations. For it is the clear consensus of the church that Copernicanism can be used to make calculations, even though the principles from which they are deduced are absolutely condemned by the judgment of the Sacred Congregation of the Index. This matter will be examined below in our reply to the seventh argument [in chapter 13]. Moreover in any event this source produces a not altogether regrettable benefit, as long as mathematicians apply their talents in a helping role, and work more and more toward trying to falsify theories rather than to defend them. But in this laborious work it will always help to remember what we quoted above in chapter 4 from the Holy Fathers, and especially the words of Ecclesiastes 3 [3:11], "He made all things good in their time, and he handed the world over to human disputes such that man will never understand the product which God has made from beginning to end."

Furthermore in using Copernicus's calculations, one can proceed in two ways. The first is to use them as purely mathematical hypotheses, on which known true physical principles are not thought to depend in any way. The second is to use them as hypotheses which are taken to be the same in kind as true natural principles, which are known to be or are reputed to be such, and from which certain demonstrated conclusions are deduced.

The first use of hypotheses is allowable as long as many phenomena are explained by the assumed theory, and as long as all periodic motions, and whatever pertains to them, are arithmetically and properly accounted for equally well by Ptolemy's principles and by the other principles used by the Copernicans. Moreover since mathematicians assume that there are lines which are infinite, or which are larger or smaller than any continuous quantity, they rightly conclude that an infinite triangle can be constructed, although the truth is that there are no infinite lines in nature. They rightly deduce that a line has length without breadth and width, that a point moves, and that a line is generated by a moving point, although this is false and physically impossible in nature. Also they rightly conclude that absolutely and simply the spheres and planets could have a greater size and greater distance from the sun located at the center. But the present order of the universe shows that it is not true that the size and distance of the stars is so immense that any star could be larger than the sun, and this is directly measured by the naked eye. Therefore, if the entire Copernican system is granted, even though it is false and contrary to reason, true calculations can still be deduced from it, and its principles, although unstable (which reveals its falsity and lack of certitude), can be applied to other matters which find their true causes in physics. For example, given the motion of the earth, someone [Galileo] has accounted for the periodic paths of sunspots in terms of the varying rotation of the earth, even though others [Scheiner] have demonstrated that this is due to other causes. And again some [Galileo] have tried to appeal, with less happy results, to the motion of the earth to account for the tides of the seas, which have already been explained long ago with many physical arguments. They actually seem to want to attribute a magnetic quality to the earth, devising all sorts of other imaginary things to account for the tides, or they say that the earth is a magnetic animal. But this is indeed a wild and a ridiculous discussion. Thus the Copernican system can be used, but only insofar as one can derive from its false premises calculations which are true and consistent with the motions of the heavens.

In the second sense of proceeding with the use of Copernicus's calculations, one is allowed to discuss its principles, but only to show that they are false, and that these calculations do not depend on true hypotheses. This is true because of the reasons stated at the beginning of this chapter. Another reason is that in many arguments, including the present one, which is based on physical and theological causes, pure mathematicians fail to distinguish between what follows with necessity because of the logical sequence and what follows from the necessity of the premises. As a result they not only reason invalidly, but they also fall into error with false and incompetent interpretations of Scripture, and they create the danger of leading others into error. This is one of the reasons why the Sacred Congregation of the Index rightly warned that the Copernican hypothesis is not to be affirmed as established, but is to be used only as a method of calculation.

The situation is different in many other cases that, although known to be matters of faith, are still debated on both sides without danger of prohibitions. For example, the existence of the world, its eternal existence, the status of future events, how generation is possible, and many other things treated by philosophers and theologians: the first cause, creation, infinity, the animation of things, the immortality of souls, and many such matters. In these latter cases it is easier to reason validly as compared to empirical topics, and it is easier to distinguish unambiguously the true from the false which sometimes seems to be true. And as a result and in contrast one finds that the firmest assent and acceptance is given to these truths of faith by all learned people, and even by those who are only moderately informed. However on topics that are not matters of faith, one can freely and unconditionally debate without any hypotheses, as long as one distinguishes between what is self-evident and what is contrary to the principles of nature.

In contrast, in the present argument concerning the motion of the earth and the state of rest of the sun, pure mathematicians test their calculations from the theory of Copernicus by appealing repeatedly to numbers, and they do not easily admit that they are reasoning from false principles. Rather they accept them as true and arrange them as postulates and axioms. And when the mathematicians reason invalidly, they err in many ways if they think that they receive any help in the matter from the theologians.

First, in cases of pure mathematics it often happens that mathematicians reason invalidly, not because of the empirical facts, as is sometimes thought, but because they have not derived their demonstrations from true causes. This has frequently happened in attempts to square the circle, to double the cube, to perpetuate motion, and in countless other cases, especially when mathematicians do not deduce necessary effects a priori but, as in the present case, they search a posteriori for imaginary causes of the effects. Just as almost all heresies in theology arise from similar invalid reasoning, likewise a great many such errors occur in mathematics. Thus [John] Gerson teaches in his *De sensu literali* that, "The true logical sense of a theological assertion does not excuse its author from the duty to reject that assertion if it is false in the literal theological sense." In a similar way it can be said that in mathematics apparent demonstrations, which are reducible to true principles, do not save the mathematician from falsity if the deduced results are false. Indeed since a mathematical problem usually requires that each step be firmly proven, it is easy to become careless and to think that a theorem has been proven and that its opposite cannot in any way be maintained.

A case in point are the many objections that are raised by various people against eccentrics and epicycles. For instance, "If the moon moves on an epicycle, the surface of the moon has a different appearance at apogee and at perigee; and thus the face of the moon presented to our vision is not always the same, which is contrary to all experience." Those who use this argument do not notice that as the moon moves on the epicycle around its own center, its motion conforms to the translational motion of the epicycle in the opposite direction, and as a consequence we always see the same spotted face of the moon. There is nothing more unusual about this motion than there is about the motion of the earth being contrary to its own diurnal motion, which the authors of that view imagine. Again those who argue that the epicycle of Venus "is so large that it passes through the center of the earth," do not distinguish between its degrees and its parts, sixty of which are contained in the semidiameter of the eccentric; nor do they designate how many semidiameters of the earth are contained in the semidiameter of the eccentric and the epicycle. If they were to substract that number from the distance of the earth from the perigee of the epicycle, they would find that the distance of the epicycle from the earth is large enough to contain easily the spheres of Mercury and the moon, and that that epicycle does not even touch the earth, much less than pass through its center. Since eccentrics and epicycles are only imaginary, then when they are used to provide an explanation of phenomena, there is much less of a problem caused if they give rise to some amount of remainder in the magnitudes under discussion than would be the case if it is claimed that the earth physically moves. The latter case would be very contrary to true philosophy, since such consequences would be inconsistent with reality. Ptolemy brings this out when he argues against the motion of the earth, to which Copernicus responds only with laughter, as Julius Caesar La Galla, among others, points out in his *De phaenomenis in orbe lunae,* chap. 7.

Many similar problems arise in the system that places the sun at the center and the earth in motion in a large orbit. This view is not only excluded by Sacred Scripture, but its contrary has been consistently verified to date, leaving aside the question of the causes of all the phenomena. Nevertheless many still are inclined to that view and defend it with apparently clear arguments, thinking that they have proven the whole matter with more firmness than can be claimed for the contrary view.

These advocates include those who claim that the conjunctions of luminous bodies are accounted for only by the motion of the earth, and not otherwise. "Given a known point of conjunction in the heavens, and given that the luminous bodies are in motion, then when the moon passes through 360 degrees of its orbit, the sun's proper motion carries it through 29 degrees or more. Hence the sun as it continues its motion is in no way contacted by the moon except at the earlier point of the full moon from which it started. Hence without the motion of the earth one cannot understand how the sun, as its degrees of motion are traversed, comes into contact with the new moon at a different point." Such is their argument.

But they clearly reason invalidly. For they take the orbits of the sun and the moon to be concentric; and when they combine the motion of the moon in the zodiac with its own proper motion, and the motion of the sun in the ecliptic with its own proper motion, they do not correctly employ the average motions as these are commonly understood. Let us indeed grant that the paths are eccentric, and that the moon is on an epicycle such that the parts of its periodic motion that seem unequal are equal by definition, and thus no inequality arises. For the lines from a true and average motion regularly fall on its center, even if the relation of orbit and center is seen from any other point in the heavens. And hence all lunar phenomena, as seen by us located in the center of the world, present themselves as at equal angles; and they always contain either equal angles, if the circles are equal, or similar and proportional angles, if the circles are unequal. Furthermore when the lines of a true and average motion differ, those lines always produce equal angles when they increase. From all this it necessarily follows that similar arcs of the eccentric and the zodiac correspond to those angles. And it also follows that there always occur at the proper time and place new moons and other lunar shapes, in regard to both conjunction and opposition, and that outside of these locations there are no other positions of the center of the epicycle and the eccentric, and no other lines of average motion of the sun. After this equality has been known for so many centuries, is it not a shame that the proportions of the motions of the sun and of the moon have not been correctly measured without use of the fiction of the earth's motion? Yes, indeed.

Even if the orbits are concentric, their arguments are not valid. For things that revolve in a circle around the same point cross similar arcs in an equal time. Therefore they intercept equal optical angles, and thus equal spaces because they are subtended by equal angles. (According to [Euclid, *Elements,*] 6, 33) they are equal, and also motions completed through them are equal. Thus it follows that the beginnings of the new moon do not touch or relate to other luminous bodies in any meaningful way, but rather occur at their own required times and places, even though in other respects the sun, which is further away, seems to move more slowly while in fact it moves faster, and thus crosses a greater arc of the ecliptic than does the moon. It is clear how defective both this argument is, as well as the other similar arguments authored by those who wish to prove the motion of the earth with necessity, and who deduce various consequences from it about the motions of the planets.

Second, in cases in which mathematicians either fear Sacred Scripture or seek approval from it, they reason invalidly and commit errors in many ways against Scripture, against the interpretation of the Holy Fathers, and against the common opinion of theologians. They begin with the following claim. "In regard to common matters about which there is no intention of teaching humans, the Scriptures speak in a human way and are to be seen as using things well-known to humans to point to divine and more sublime matters." In this view they are wrong first when they suppose that, in regard to the motion of the sun and the earth at rest and similar things, it is not the business of Scripture to teach humans. This is openly contrary to Ecclesiastes 1 and to Ecclesiasticus 24 and 43. Furthermore Moses would have written the history of creation in vain if he had not wished to direct the instruction of humans in faith, in morals, and in science. For historical instruction has no other purpose than to see the morals grounded in it. Finally they err because they do not see that it is one thing to speak to humans in a human way, and quite another thing to lead them away from falsity and to a knowledge of the truth. For it is clear that divine and more sublime things are signified by matters which are well known to humans, either by parables or by similitudes or in any other way. Nevertheless the Sacred Scriptures never use a similitude or teach in any way by starting with what is in itself false or contrary to the senses or to human opinion. Hence these good mathematicians are in error when they claim, from their view stated above, that, "It is not surprising that the Sacred Scriptures speak of sensible matters, to both learned and ignorant humans, even though the truth of the matter is contrary to these sensible matters." It will be made clear below in the final chapter how Moses used sensible appearances in addressing ignorant people.

Furthermore they try to confirm their view by saying, "Everyone knows that Psalm 18 [19:4−6] makes a poetical allusion when it uses the image of the sun as it sings about the path of the Church and about the journey of Christ triumphant in the world." But they do not properly distinguish between the mystical sense and the literal sense on which it is based; for if the literal sense is not true, then it does not rightly signify the mystical sense. When they claim that "poetry was primary for the Hebrews," they fall into trivia. As a result they have not made their point that literal truth has not been spoken by Moses, or David or Solomon or any other scriptural writer, while it has been spoken by Musaeus, Linus, Orpheus, Hesiod, Homer, and other secular writers whom they treat as divine. Finally they are doing theology on their own when they add that this passage of Scripture is imitated by Virgil's "Aurora was rising from Tithonus' golden marriage bed" [*Aeneid,* bk. 4, l. 585].

Third they claim, "The sun does not depart from the horizon as from a tabernacle, even though that is how it appears to the eyes; and this was known to the Psalmist, who still maintained that the sun moves, because that is how it appears to our vision." But they do not realize that they thus impose on the Scripture and the Psalmist the view that the literal sense is based on false appearances. For the proper truth instituted mysteriously by the Holy Spirit cannot, and should not, be based on optical illusions or imaginary notions. This would make the true to be false, and apparent only in dreams. That truth is not based on a literal truth that is contrary to the tradition of the Scriptures, which would conceal the gospel and the path of the Son of God on earth. Further, when they add that "David's intention was to display the great and clear magnificence of God, who caused the sun to be so represented to our vision," they reason invalidly again. They do not understand that David did not carry out his own intention, but the Holy Spirit's, since he perhaps did not even know what mystery was contained in the words he pronounced. For as they speak by divine instigation and feeling, it is not necessary that the prophets either witness the matter itself or recognize it in their symbols. The same is true of their imaginary account of Joshua ordering the sun to stand still in the middle of the heavens.

They do not notice that they have imprudently argued that God is the author of a falsehood, when they try to base the truth of faith on an apparent reality and the mystical sense on a visual illusion and a false appearance.

Fourth, another interpretation is that "the sun stood still" means "the earth stood still," a view which some say is an absurd result of the Copernican hypothesis. Its defenders say that these critics "do not realize that this is only a verbal opposition within the confines of optics and astronomy, and it has no effect outside in practical human affairs." But they do not realize how they contradict themselves and true philosophy. For if they maintain that the stability of the earth and the motion of the sun are only matters of appearances and of human sensation, then the opposite is surely located in the external world and it ought to affect human affairs. This is especially so if, contrary to the common philosophy, particular sensory illusions occur with a common sensible like motion, which cannot be corrected either by the senses themselves (which have a fixed object), nor by reason, unless reason thinks up many absurdities which have never come under the senses, as happens in the present case. For if someone deduces by reason that the sun stands still, he conceives of the earth in motion, and his reason must form all those monstrosities that the defenders of such a motion think must be imagined and accepted as true, for example, that the heavens are at rest, that the stars of the firmament are separated from each other by truly immense distances, that heavy things do not tend toward the center of the universe, that bodies thrown upwards do not fall on the perpendicular but are moved with the air in a circle, and innumerable other such things.

Therefore if the sun seems to move because the earth truly moves, then the latter is the sole and basic cause of the apparent motion of the sun. There is no inconsistency within the field of optics unless it also exists within the confines of nature, and that it would be true to say that "the sun stood still" means "the earth stood still." Or if one simply reverses the logic, the question arises as to how causal hypotheses apply to "the sun shines, it is daytime," and other such cases presented in the same way, as either interpretive or as causal and inferential, or as a simple conversion. Otherwise there will be no agreement between physics and optics, which no one will rightly claim.

Moreover, if the apparent motion of the sun is relative to the true motion of the earth, then it follows that when the earth is taken at rest, the sun must be seen to be moved, although it is known to be truly at rest. It also follows, on the other hand, that the stability of the earth is investigated by using the stability of the sun as a symbol; and this happens by reason or by sense or by both. If by reason, we know that the parallax relative to the fixed sun is the same everywhere. The parallax, when the earth is moved as before and the sun by hypothesis is truly at rest, is the same as when the earth is taken to be at rest, although we ought to judge by reason that it is moved, as the terms imply. Or if the sameness of parallax is referred to the apparent motion of the sun as before, in that event the apparent motion of the sun also will remain invariant to the senses, because the parallax is exactly the same. Therefore when the earth is taken to be at rest and the sun apparently moves, the physical cause of this apparent motion must be something other than the motion of the earth. And thus the whole philosophy and astronomy of Copernicus collapses. Otherwise, if the translation of the earth were the sole cause of the motion of the sun to the senses, and if the parallaxes are still the same when the earth is taken to be at rest, then it follows in the given case that the earth both stands still and moves, so that, given this, the sun also seems to move but nevertheless is truly at rest.

Now if the earth is perceived by the senses to be at rest, there is no reason why it is not also perceived by the same senses at another time to be in motion. But this cannot happen according to the adversaries' view, namely, they say that the earth is judged to be at rest only apparently and relative to the senses. It is not possible to give a reason why the motion of the sun is taken to be apparent because of the combined apparent rest and true motion of the earth, rather than the contrary that the motion of the earth is taken to be apparent because of the combined apparent motion and true state of rest of the sun, as they say. They allege that, "The sun moves apparently, but not the earth," because, "since the sun appears to be small and earth very large, the motion of the sun is not comprehended by vision because of its slowness, but only by reason because of its changing closeness over time to the mountains." But this does not answer the difficulty. For even if there were no mountains, the motion of the sun would still be perceived after a greater and greater interval is produced from any fixed point, real or imaginary, and certainly not because of the motion of the earth. For the same thing would happen even if the earth were not located in the center, and if in place of the center there were an imaginary point or an eye observing that motion. Therefore, as with what must be said about the motion of the earth being apprehended relative to some fixed point, which cannot happen without an imaginary and false state of rest of the sun, this whole philosophy is false and imaginary. Hence it cannot be in agreement with optics or physics.

Finally if the stability of the earth in the given case is examined by both sense and reason, this does not mean that the stability of the sun is perceived in both ways, but rather that the true motion of the earth would be recognized from the apparent motion of the sun, or that the apparent stability of the earth is clearly recognized from the true stability of the sun. For the apparent stability of the sun is related by sense and reason to its apparent motion in the same way that the true stability of the earth is related to its true motion. Therefore if we commute the proportion, the true motion of the earth can be perceived from the apparent stability of the sun, and the true stability of the earth can be perceived from the apparent motion of the sun. Or if the proportion is treated as a homology, then the apparent stability of the earth is inferred from the true stability of the sun, which can be joined with the true motion of the earth. In that case the earth indeed stands still only apparently, but is physically moved. And on the other hand the sun in reality truly did stand still when it was ordered to stop from its true motion. But the opposite view would greatly disturb all things, which indeed destroys Copernican philosophy. It is better to say in this event that the earth has always stood fixed previously, that it has no relation to the motion or to the stability of the sun, and that the whole miracle in the sun occurred both really and apparently.

This argument will come up again later. Here we attend only to the following. If we wish to be the least offensive in using the Copernican theory, which is permitted in the way described above, then we should act by always keeping firmly in mind that the truth is opposed to the principles from which this theory is deduced, that nothing has been demonstrated by them, that there are no proofs contrary to truth and especially to the content of faith, and that true arguments based on exact calculations are only hypothetically deduced from the force of the consequence, and not from the consequent of the theory. However one must be very careful not to slide from one category to the other, that is, from the deduction of calculations to an argument for the principles or positions as true. For this reason they do not give good advice to the community who verbally broadcast the rules considered here and used by learned scholars, and who try to notify indiscriminately all those who are idle and dishonest. Indeed such behavior and knowledge should be abandoned if it is harmful to all the more important disciplines, especially those that deal with the more abstruse doctrines of the faith.

Finally it should be observed that, in regard to the physical principles of Copernicus, the contrary true view must always be expressly substituted or established. And one should always keep in mind what was said by Augustine in his commentary on Genesis in *De Genesi ad litteram,* bk. 2, chap. 5, "The entire power of the human mind is less than the authority of Scripture," or as I would say for the present case, "is less than the authority of the faith."

Also, in session eight of the [Fifth] Lateran Council under Leo X, in a similar vein it is ordered that, when the principles or conclusions of the philosophers deviate from the true faith, then the teachers of philosophy are obligated to correct them with the manifest truth of the Christian faith and to refute the arguments of the philosophers. The words of the decree are as follows. "For each and every philosopher lecturing in the universities of general studies or elsewhere in public, we issue the specific rule that, when they read or explain to their audiences principles or conclusions of philosophy which are known to deviate from the true faith, such as the mortality or unity of all souls, or the eternity of the world, or other such things, they must make every effort to make clear to their audiences the truth of the Christian religion, to teach that as persuasively as possible, and to make every effort to refute and reject vigorously the arguments of such philosophers with every available response." So said the Council.

Note in the above words that after the cases given which are expressly condemned "as assertions contrary to the truth illuminated by the faith and utterly false," there is a general extension of the order "to other such things." This could include the assertion of the motion of the earth and the stability of the sun, either as involved in the eternity of the world and its motion, or as simply deduced from principles contrary to the truth illuminated by faith (as we have tried to show up to this point). Hence that opinion seems to be included and condemned along with other assertions. For in extending an order it is enough if the same rationale behind the decision applies, even if in some respect the rationale is weaker in the case to which it is extended, as is stated in *Caput translato, extra de constitutionibus.* Panormitanus [Niccolò de' Tudeschi] also says in the same place that "everything inherent in these assertions contains the same error" as "destructive of the Christian faith." We will pass over the other things which the above mentioned decree says should be "avoided and punished."

Furthermore the decree permits an explanation with similar principles and the refutation of contrary arguments, provided that the truth of the Christian religion truly makes that evident. For the principles of philosophy are never properly presented unless they teach how pagan superstitions are destroyed as false by the truth of the faith. This was especially obvious when the philosophy of Aristotle was introduced into the schools, for what he said was hardly ever better explained than by refuting the things that he thought wrongly. But this point does not apply to the present argument about the motion of the earth and the stability of the sun. For this Pythagorean dream has never been received into the schools, nor has it had any impact on religion. Therefore the Copernican theory and its related Pythagorean philosophy should not be taught at all; neither to explain in the schools its physical principles, from which a dispute might arise leading to the need to firmly establish the contrary as true; nor to simply reject it by showing its falsity or by completely outlawing it by principles of physics and also of theology.

As a result, if the physical principles of the Copernican theory are clearly rejected and destroyed, as we have said, then the truth signified by the contrary view is either granted or established, and every occasion of error is avoided. Otherwise we would not exclude from mathematics what the Pyrrhonists say in philosophy, namely, never decide on anything and doubt all things. Or otherwise, if this theory is constructed contrary to the principles of the faith, then likewise we should favor Protagoras of Abdera for not being able to say how the world either did or did not begin. Or otherwise we should agree with Diagora and Theodorus of Athens about the constitution of the world, and thus we would find fault with the wisdom of God the creator and remove God from the center of things.

For this reason what the Christian philosophers and the more holy theologians do in their fields ought to serve as an example for pious mathematicians in their work. Clearly when St. Thomas raises the question of the existence of God in his [Summa theologica,] part I, q. 2, art. 3, he does not try to discuss any specific arguments that God does not exist, but rather, leaving the faith aside, he shows with many arguments that God does exist. And in article 2 he teaches that that can be demonstrated from God's effects. And he uses the same method when he undertakes his discussion of questions relating to the faith and to the doctrine of morals. Another clear example from the beginning of the Church is the practice of the Holy Fathers when they raise objections against pagans and heretics. And today there is no other method used in the schools and academies of the wise. Without doubt the contrary is the case for those who reason invalidly by absolutely affirming the motion of the earth and the stability of the sun in the center, thereby neglecting or even attacking the truth of the opposite view, and not attempting to inquire into and to establish the truth more deeply.

CHAPTER 13

Refutation of arguments militantly in favor of the motion of the earth

We turn now to arguments that make an appeal to Sacred Scripture. A few of these arguments do that because without religion their own teaching would seem to be unprotected. Although they honor scriptural passages themselves for their meaning, they almost always use them wrongly. Many other arguments distort the passages of Scripture, either by advocating what is opposed to its truth, by maintaining additional meanings that are not thought to belong to Scripture, or finally by using fallacies rather than adhering to Scripture, which is almost what the first group of arguments does.

FIRST ARGUMENT

The main point is that the passages of Scripture that refer specifically to the rest, stability, or firmness of the earth, or of the orb of the earth, are correctly understood to refer to the position and order that the moving earth preserves as firm and perpetual for a definite time. And therefore just as the heavens are called the "firmament" because, although they move, they preserve the same firmness and constancy in their motion, so also, although the earth revolves in a circle, it retains the same site in respect to itself and the same order in respect to its motion. The text of Ecclesiastes 1 [1:4], "The earth stands firm forever," is to be understood as referring to its state of being indestructible, especially in contrast to humans, who "never remain in the same state," Job 14 [14:5–12]. What is said in the same place [Ecclesiastes 1:5] about the rising and setting of the sun is to be understood according to the senses and appearances. For that reason St. Thomas says in [Summa theologiae,] part I, q. 70, art. 1, ad 3, that Moses speaks of both insofar as they appear to the senses. In short there is not just one interpretation of this passage, but another one follows easily from the teaching of Ptolemy and Aristotle concerning the earth at rest and the sun in motion. Let us add that among the authors there is not just one explanation of the rising and the setting of the sun. Along with others St. Augustine interprets the rising and setting of the sun as its being first hidden behind the northern mountains and its later emergence, without its revolving around underneath the hemisphere of the earth. Therefore, etc.

RESPONSE: I deny the antecedent in regard to the site and order of the earth's motion, but concede it in regard to the earth's order and its absolute site, in which sense the power of nature has given it an invariable position in the middle of the universe which we have said elsewhere is the interpretation of St. Thomas in *Opusculum 10*, art. 16. And therefore the earth possesses a stability and firmness without motion, which is indicated here and there in the Scriptures. Regarding the argument I say that the heavens and the earth do not have the same type of firmness. For although the heavens move, they do not change the order of the universe, but the motion of the earth would change it, especially if the earth exists among the planets and the sun exists at the center. Even if the earth were to rotate circularly in its own place, a change of the order would not result. The earth's motion in any other way would be contrary to its nature as established at creation, for resting naturally at the center follows from its purpose.

Regarding Ecclesiastes 1 [1:4], that passage in Scripture is not taken only in that sense, as we have discussed at length in chapters 5, 6, and 7.

Regarding the other passage about the rising and setting of the sun, I respond first that, in addition to what we said in chapter 9 from St. Augustine and Gregory Nazianzenus, it is better to say that the sun and the stars stand still according to sensory appearances. For unless a period of time passes when one perceives points approaching or departing from the earth's axes, which turn more slowly among the stars, these points clearly seem to be fixed in the firmament. As a result the Greeks gave them the name "aplanes" [non-wandering]. The same thing happens on the earth. For objects located a great distance from the viewer will be seen from a narrow angle as standing still, even though they are moved by the same cause, and this happens as long as the distance is unchanged.

I respond secondly that if the sun moves according to the senses, then the earth does not move according to the senses. This itself is sufficient to prove that the earth has a fixed position. For the senses do not judge that the sun moves unless they perceive points approaching towards or receding from each other. But they do not perceive such points in respect to the earth, for the earth always maintains the same longitude and latitude. And this same invariance of the poles rightly indicates that the earth is at rest and that the sun moves.

In response thirdly I deny that the rising and setting of the sun can be dismissed as merely sensory appearances, as if they do not exist in reality. Otherwise many passages in Sacred Scripture would be placed in doubt. For example, the miracles performed by Moses before the Pharaoh would have to be illusions and deceptions, and not miracles. And the same would have to be said of many cases, and in some cases about what happened both in the wilderness and elsewhere, it would be sinful to say that they occurred only according to the senses. It is truly amazing that almost all of the Holy Fathers understood the motion of the sun, and not one of them thought that this was only an appearance of the senses, as these authors themselves have imagined.

There is an objection from St. Thomas to the effect that it is not clear whether Moses spoke of these things as they appear to the senses. For in the place quoted above [*Summa theologiae,* part I, q. 70, art. 1] St. Thomas raises the question of whether the luminous bodies ought to have been produced on the fourth day, when in the third objection it is urged that they ought to have been produced on the second day with the firmament to which they are attached. He responds, "According to Ptolemy the luminous bodies are not attached to the spheres, but have a motion different from the motion of the spheres. According to Aristotle they indeed are attached to the spheres, and have no motion other than the motion of the spheres. According to the truth of the matter however, the motions of the luminous bodies, but not the motions of the spheres, are perceived by the senses. However Moses deigned to speak to uneducated people by dealing with things which are apparent to the senses." It is as if St. Thomas had said that Moses did not talk about the creation of the luminous bodies on the second day along with the firmament in which they are fixed, although he did talk about the creation of plants with the formation of the earth to which they are attached, because uneducated people would not have understood the distinction of nature made on the second day. For the senses perceive the distinction made on the fourth day in which the sun was clearly seen to move by the senses, even though they would not have discerned the motion of the firmament. From this it obviously follows both that a true motion of the luminous bodies is perceived by the senses, and that Moses knew that it was easily perceived by the senses. But he remained silent about the motion of the firmament, which uneducated people could not readily discern, whether the stars were attached to them or not. St. Thomas gives another response in the same place, but this is sufficient for our purposes.

It should also be noted that most of the Holy Fathers say that in Genesis Moses spoke according to the senses and the capacity of uneducated people. But this should not be understood to mean that what is said about sensible things are only appearances and not the truth. Rather since there are many hidden truths which are far removed from the people, and which are not understood by the listeners, it would be easy to lead them into error. And thus one should not infer that what is said about such things is not true in respect to nature and the intellect, but are just appearances to the senses.

Thus St. Chrysostom in his *Homily 2 on Genesis* explicitly omitted a discussion of spiritual creatures, lest uneducated people, who can grasp only physical matters, be seduced into worshiping them as Gods. And this is implicitly said by Augustine in his *De civitatae Dei*, bk. 11, chaps. 9 and 33, where he insinuates that the words, "In the beginning God created heaven and earth," are captured by the word "heaven." He says the same thing about, "He divided the waters from the waters." He explicitly mentions water and earth, but does not use the word "air," not because "air" would not be true, but because it is something unknown to the uneducated beyond what is clearly apparent to the senses. Nevertheless when he says, "Darkness was above the surface of the abyss," he signifies to the educated that the air itself, as connected with water, is certainly a diaphanous body above the surface of the water and that it is affected by light and darkness. Furthermore (as we have seen) he puts the creation of plants before the creation of luminous bodies in order to exclude them, as Basil says in his *Homilia 5 in Hexameron idololatriam*, "Those who believe that the luminous bodies are Gods say that plants have their very first origin before the luminous bodies."

Finally when the Bible speaks of the heavens and the stars, it does not discuss their motions; nor whether they move themselves and thus are thought to be endowed with souls; nor does it assign them any purpose other than what would be seen to be useful for the people. For as Deuteronomy 4 [4:19] prohibits, "Do not raise your eyes to the heavens and see the sun and the moon and all the stars, and be led into the error of adoring things that the Lord God created for the use of all peoples under the heavens," such as the Gentiles, who are talked about in Wisdom 19. And in the Christian era this error of commanding that the moon and the stars be adored has dominated some sects, especially the Manichaeans and the Marcionites according to Constantine of Harmenopolis in his book *De opinionibus haereticis*. Furthermore Moses has explained that God's purpose in creating them was threefold: to give light to the firmament and to illuminate the earth; to establish the days and years of time; and finally to establish signs. But the Bible does not openly explain how these things happen, leaving their explanations to others. Indeed Ecclesiastes 1 accounts for them through the various motions, approaches, and recessions of the luminous bodies as signifying all things that happen. The same things are signified from time to time in various other passages of Sacred Scripture, which many Christian philosophers have then chosen to quote.

In these passages it is clear that Moses never spoke about what was sensed by the people unless that meaning was based on the truth, even though he had not perceived all things. And moreover he was not silent about true but more subtle matters which he would insinuate without any explanation to those who were more capable of understanding, or he would leave the explanation for another time and place. On this see St. Thomas [*Summa theologiae*], part I, q. 67, art. 4; q. 68, art. 3; q. 69, art. 2, ad 3; and q. 70, art. 1, ad 4.

Finally it should be added that the words of Ecclesiastes 1 concerning the rise and setting of the sun do not have just one, but many, explanations. It is especially easy to admit this if one also admits the opinion of Epicurus in *Cleomedis* [*Caelestia*], bk. 2, "The sun obviously dies and is born again." Here also this saying about the rising and setting of the sun would be considered to be ridiculous as understood by uneducated people. Or if you think like the poets, the sun immerses its leading edge into the water of the ocean, as found in St. Augustine, *De Genesi ad litteram*, bk. 1, chap. 10, and in Strabo, *Geographia*, bk. 1. However we are not interested in fables, but with what we have shown in chapters 8 and 9 to be holy and true in the literal sense. Then there is the point that St. Augustine elsewhere did not want to say that the sun rotated below the earth, but that it was hidden behind the mountains on the northern side of the earth. However I know from what he says in *De Genesi ad litteram imperfectus*, chaps. 12 and 13, and in *De Genesi ad litteram*, bk. 2, chaps. 13 and 14, that he very early abandoned the view that the sun does not revolve around the earth. And indeed in *De Genesi ad litteram* bk. 1, chap. 10, he clearly holds the opposite view, although some see in that passage a lack of thoughtfulness, while others think that is not so. Nor is there any objection to the interpretation of St. Thomas in *Opusculum 10*, art. 10, and in *Opusculum 11*, art. 6, about an angelic spirit moving the sun. For this is not a reason why it would not move in a circle, and thus it could be said to rise and set and to return to its place, and in one word, to move, which is sufficient for our purposes.

SECOND ARGUMENT

The starry heaven is called the firmament only because it is at rest and does not move. Therefore it is true that the earth moves. The antecedent is obvious. The

Letter to the Hebrews 8 [8:2] uses the word "heavens" to mean "the Tabernacle which God established." And chapter 12 [12:28] speaks of the "immobile kingdom." Chrysostom says the same thing both in his Homily to the People, *6 and 13, and elsewhere. And St. Augustine says in* De Genesi ad litteram, *bk. 2, chap. 10, that the mathematicians of his time had demonstrated this. The consequent is proven as follows. Since the phenomena which we continually experience can be saved in no other way, then it is self-evident. Furthermore this clearly follows from the miraculous events in Joshua 10 [10:12], where it is said that the sun stood still in the middle of the heavens. And Isaiah 38 [38:7] says that the sun moved backwards as a sign of Hezekiah's health, and so the sun truly had stopped and moved backwards. But these miracles did not happen to the sun, although the Scripture says that because of our senses. 3 Esdras 4 [1Esdras 4:34] says, "The earth is large, the heavens are lofty, and the course of the sun is very fast as it rotates around the heavens in its proper place in one day." This passage must be understood to refer to either a true motion of the sun through the poles of the heavens; or else to its apparent motion; or else to the sun's light, which rotates the planets without any motion of the spheres in which they are embedded; or else that the sun affects the air and vapor which results in a circle of vapors around the planets; or finally to a circular rotation of the sun on itself in its own place. This last alternative is in fact how the very fast motion of the sun occurs. Therefore, etc.*

RESPONSE: I deny the antecedent. For as St. Augustine teaches in *De Genesi ad litteram*, bk. 2, chap. 10, "The word 'firmament' does not mean that we should think that the heavens stand still, for the heavens are said to be the 'firmament' not because of their being at rest, but because of their firmness or because of their being an impassable boundary between the upper and the lower waters." Further, whatever the nature of these waters may be, they are like this according to the various opinions of the Holy Fathers.

Regarding the proof of the antecedent, I respond that in an earlier passage St. Paul took the word "tabernacle" figuratively, and St. Augustine did not even once use the word "vault" for it. The word "established" refers only to the firmness of the boundaries between the waters, as having an invariable law in respect to its purpose, in relation to both the whole heavens and to the parts which compose it, whether the heavens be solid or liquid, which makes little difference here. In a later passage Paul says that the heavens are an immutable, that is, a permanent, kingdom. It is never mutable in relation to the state of beatitude, as compared to the world and the state of being pilgrims, in which we do not have the permanent citizenship which we seek in the future.

St. Chrysostom was also of the opinion that the heavens in themselves do not move, although he said the planets and the stars revolve daily. This is enough to say that in his opinion the earth does not move, which would thus refute this second argument. The planets and the stars revolve daily, and hence the earth does not move but is at rest. For it is necessary for something to be unmoved around which the planets and stars revolve, and nothing besides the earth can be assigned that role.

In the passage quoted above St. Augustine does not affirm that the mathematicians of his day had demonstrated that the firmament is at rest. He says this only hypothetically, for after he said, "firmament not because of their being at rest but because of their firmness," etc., he added, "If the heavens were truly shown to be at rest, we would not be hindered by the circuit of the stars, and not be unable to understand this. And those who have inquired into this most carefully would have found also that, if only the stars revolve and the heavens are unmoved, then everything that would have happened would have been noticed and understood in these rotations of the stars." These remarks by Augustine are very far removed from the thrust of the argument. Moreover the most careful investigators have not asked whether the heavenly planets and stars are all immobile, but only specifically whether the earth revolves and would everything then happen that we actually see happen. Or perhaps like the Pythagoreans they asked but found no answer.

Furthermore it is quite clear from the thinking of St. Augustine that, even if the antecedent is granted, the consequent, namely, it is not possible to save the phenomena unless the earth moves and the firmament is motionless, is false. Indeed if the heavens do not move, and if only the planets revolve in their own periods, this also must be related to the phenomena, according to the thinking of the mathematicians of that time. And if both move, then it must be asked whether all the same things that occur every day are plainly seen. Quite to the contrary, if the heavens and the planets do not move and the earth alone revolves, then many things, which are seen as fixed in place and in duration, could not be seen at all. And if we assume that the heavens are not vaulted but are flat in their extension, as some of the ancients thought, then when one looks directly at appearances, the radius in the usual sense would make either a right angle or an oblique angle with the radius coming from the center of whatever is seen. And if we assume that the heavens are spherical (which is the case), then, unless the earth is agitated by very many different types of motion, it would be moved toward all axes, and it would either wander widely here and there outside the ecliptic between the tropics in its own circular motion, or it would return in a spiral. In either case very many phenomena that we observe every day would necessarily be hidden. All these things are obvious to anyone who gives it the least attention. But taking that into consideration, no one would recommend so many different types of motion, although elsewhere one might accept a strange motion of the earth.

What is said about those prodigious events which occurred in the times of Joshua and Hezekiah about the earth being at rest and the miracle of regression is devoid of truth. For otherwise the earlier event, which relates to the miracle of Joshua, would not be correctly described in Scripture. "And the moon over the valley of Aijalon" [Joshua 10:12] would not mean that it was stationary. For according to the authors of the argument the moon moves in a small orbit in a proportionally slow motion, and the earth in a large orbit, from east to west in succession. Hence if the earth did not come to a stop, then the moon could no more be said to be stationary above Aijalon, which was east of Gibeon, than that it was stationary above Gibeon. For in the meanwhile the moon would have had to revolve in its own path and motion, as indicated, lest it approach the sun, which must have gone on ahead. This would have thus shortened the day of the battle, whether the sun was above the equator or not, and the moon would have fallen down outside the solar system. Otherwise the order that the sun stand still would be vacuous. And then more importantly we could say with others that the moon was in the east, and therefore was full, or close to being full, at the beginning of the third month. In that case the moon had neither moved opposite to the sun, nor had it reversed the direction of its motion. Therefore there was no reason for the moon to be ordered to stand still, or to move above Aijalon which was already to the east.

If they wish to say that the moon also is by nature at rest, and that only the motion of the earth was interrupted for a time, then as a result the order and the respective location of all the stars would be destroyed. This would introduce an extraordinary confusion in what is said in the Scriptures, which would be inexplicable by any rules or figures of speech. This would also be clearly contrary to reason, for when it is said that the sun or moon moves, we would believe that they are actually at rest, and we should be persuaded contrary to our senses that the earth moves. And on the other hand when the sun or moon was ordered not to move, we are to understand that the earth, by nature immobile, continued on its course. To this we add that in the future it would not be less of a miracle if the earth, or at least the greater part of it, which is a huge and sensible mass, would stand still, held rigidly by a human voice. For indeed this particular miracle ought to have remained as a constant reminder of the greater glory of God for those for whom it was done. But for those whom no sensible evidence of the motion of the earth has persuaded that this could happen, and who hold rather the contrary to be established, such a miracle would produce laughter rather than the glorification of God. Clearly the command was for the earth to stop, although it is naturally at rest and is motionless to the senses even more than to reason. This miracle taught a lesson to the people who were not watching that they necessarily ought to perceive, unless we say now that the Pythagorean system had been proposed to and explained to everyone. And thus in the brief interval after Moses, when Joshua lived, the latter nevertheless followed the example of Moses, and spoke to the senses of uneducated people that the miracle occurred in the sun and the moon, perhaps because it could not have happened, as they had seen on the earth. But these are mere trifles.

Nevertheless the authors themselves claim support from the following line of Virgil: "We move forward from the port, and the earth and cities recede" [*Aeneid*, bk. 3, l. 72]. "The sailors do not perceive the motion of the ship, but do think that the beaches, earth, and cities move, even though otherwise they are motionless. Likewise the earth moves, even though we do not sense how very fast it moves, and as a result the sun, moon, and stars are thought to revolve, even though otherwise they are motionless." So these authors say. But this argument is silly. This experience is more easily found in the game of vertigo (commonly called "fly" from Aeneas), in which to a child who rotates around by twisting in the same location but not in the same orientation, all things seem to move in any direction because of the motion of the child's breathing. In the present case experience shows that, if the sailors do not keep their gaze motionless on the ship which carries them, they will relate their own progression to the ship which carries them, and physical bodies will be sensed as moving away behind them in a slow motion as if they were moved by a true motion. Because of the motion of the eyeball, it happens that they are seen as being carried past the eye, and become converted to things left behind, even though they are known to be truly at rest. But we experience none of this in regard to the earth. For even if the earth were to move very fast and with many motions, the planets and other stars meanwhile are still very remote, and we would notice that they actually move or are moved, which we have proven with various arguments. Thus when we sense them over distinct periods of time, whether the eye be moved or motionless, we on earth have absolutely none of the experiences which would occur on a moving ship.

Regarding the motion of the ship, it could be added either that a chariot carrying physical objects does not seem to move faster than the ship itself, or else that such a chariot moving in the opposite direction would be thought by the rules of optics to move twice as fast. However, although we have seen that the heavenly lights move with a very great speed, we do not have any experience of the true motion of the earth. And an imaginary motion would vary in relation to the motion of the ship. For if the ship goes west, it will seem to go east, and to the contrary, if it goes east, it will be thought to go west, and there are other distinctions concerning the earth itself about which our authors are unable to agree. I have said that the heavenly bodies seem to move very fast, both because of their size no less than their immense distance, insofar as they are studied from an optical point of view, other things being equal. Thus to the eyes which freely follow their motion, they are seen to move almost like streams of water gliding by when inspected more attentively, and are also perceived to run downward like a wave of water when seen rapidly by an intense glance. Moreover, to speak specifically of the stars in the firmament, even though they move with the fastest speed and across the longest space, the angles seen between them are minute and they seem to be at rest rather than in motion. Nevertheless after time they are perceived to have changed to a place very nearby. The same thing is true of objects seen in a middle position and with a common radius stretched to the horizon. This happens both when straight lines depart from the eye, and when their path cuts across the straight line radius.

Regarding the later miracle about what happened to Hezekiah, the falsity of the argument is clear from what was said above in chapter 8. To that we add that King Hezekiah was not so uneducated that he knew nothing about the motion of the earth, and that he would have had to ask about the miracle implied in Isaiah's words. He certainly would have been able to grasp Isaiah's message in just one sentence, "I wish the earth would move backwards," and Isaiah could easily have stated the same thing with, "The earth has turned backwards," and not with, "The sun has turned backwards." For in the palace of the king, or rather in the whole kingdom, the motion of the earth could have been understood. This was certainly the case because so many centuries had elapsed since the time of Moses, and by then the people were no longer so uneducated that they would have perceived only what was apparent to the senses and not what was proven by reason. For who would doubt that Hezekiah, and all the thoughtful people of his time, would have previously read what had been written in Ecclesiastes about the motion of the sun and the state of rest of the earth, that they would have experienced the same things daily, and that this had been proven by rational arguments just as described in Isaiah's statement? For he clearly had spoken about the retrogression of the sun, not the retrogression of the earth, and nothing would have been easier to occur, if that is what actually happened.

To this we can add that, although Scripture in some passages speaks in metaphors and figures of speech, and in other places it speaks explicitly and properly, it is truly remarkable that it nowhere clearly says that the earth moves or that the sun is at rest. These things would be great miracles of nature, large enough to establish the knowledge and glory of God, like the many things mentioned in Wisdom 13; yet quite to the contrary they are not clearly contained in, or even implied in, a single passage. For uneducated people never persevere in their beliefs about anything unless it was grasped by the senses, especially if it is something that refers to God for their admiration. For who on earth would not say that the revolution of the earth and its waters, leaving out of consideration all of the living things in them, is a greater miracle than the present motion of the sun, even though it cannot be decided whether the sun is firm or moves in a liquid or a solid space? In this one argument all the dangers of a later idolatry are present from the beginning. For if the sun and stars lack motion and have no souls, even though they seem to move, then people would have thought that the cause of these appearances is the rising motion of the earth, lest they be ridiculed for a long time by various objections. Or else, as we have already said, Scripture would have to be taken in the opposite sense, so that when it is said that the earth is at rest and the sun moves, we should understand that to mean that the sun is at rest and the earth moves.

Moreover, as is said in Wisdom 19 [19:18], each of the elements contributes its own sound to maintain a harmony of sites, orders, and other roles, as was said above from Clement of Alexandria in chapter 4. Just as it is true that a proportion is presumed among all of them, this also includes the earth because of its state of rest. From this it is possible to infer specifically what is added in the same passage. "This can be truly grasped from seeing it," that is, the earth is at rest and is in no way moved without any discussion or perception needed by uneducated people. On the contrary this can be attributed to the celestial harmony mentioned in Job 38 [38:7], "The celestial harmony puts one to sleep." For the sun, which is indeed a special part of this harmony, is a string of the celestial musical instrument which continually moves and is never at rest, and can be truly grasped from seeing it.

Finally what was added in the argument to explain the passage from Esdras is ridiculous. For if the sun is at rest, and if it sends out rays and light equally in all directions, then there is no reason why it would turn the planets in one direction rather than another, or why it could control their motions while it itself has no motion. Secondly for the same reason the defenders will see that just as it is not coherent to say that the motions of the planets interfere with each other, the same is also true of both the motion of the earth and the motion of the great sphere in which the earth is carried. Thirdly what is said about the air and vapor moving around the planets is not defendable. For if they are moved by the sun to move the planets, the first difficulty returns. And if they are moved by a source external to and distinct from the sun, then it can be said that the planets move either by themselves or by the motion of their spheres, and thus they have no motion from the sun. And indeed the same thing applies to apparent motions, and thus all is repeatedly thrown into disorder.

The idea that the sun revolves on its own axis is not helpful. For if the sun has a uniform motion in one direction, it ought to carry the planets with it in the same direction, so that they are rotated in one day to the same place in the circle. And it can hardly be explained how the planets can move in one direction and the sun in another direction, and especially how the moon moves in a small circle around the earth (the same argument applies to the satellites of Jupiter in their orbits around Jupiter). For if the moon rotates in its own orbit each day with a motion that is independent of the earth's motion, then a new moon and a full moon would occur each day. Or if it continues on with its own regular motion each day, it would not derive its daily motion from the rotation of the sun, unless it were moved together with the earth by the motion of the earth's great sphere, which those who deny the spheres would not be willing to admit. Others would not be bold enough to admit this, unless they could determine that it previously had a regular or irregular rotational motion, which I doubt they could do.

The same difficulty arises in regard to the earth's proper motion, which occurs annually around its center, and which is contrary to its daily motion. For it is not possible to understand how it could move contrary to the revolution of the sun. For if the motion of the sun were to cease, who could ever doubt that the planets and the earth would stop moving, for given the nature of the cause, these effects depend on the sun's actual influence? In a like manner, when Joshua ordered the sun to stop, then the cessation of the sun's rotation on itself can be understood in the literal sense. In regard to the meaning of the Scripture text, "Sun, do not move over Gibeon," or as the Septuagint says, "toward the west," I say the meaning will be, "Sun do not move toward the east," which is the direction of rotation. Or if the planets ceased only from their daily motion when the sun stopped, and continued on their annual motion, then it would be false to say that the earth was at rest. And the miracle did not occur to the sun but to the earth, since the earth was only apparently at rest.

As a result the meaning of the words of Esdras is not what the above mentioned authors imagined; namely, that the sun actively turns the heavens, or to put it in another way, that the planets or the heavens are passively turned by the sun's motion or by its power. Rather the meaning is intransitive, as is signified by the words "turns," "goes through," "completes in its circle" the whole sphere of the heavens, and returns "to its own place in one day." The latter words refer to the diurnal motion in order to exaggerate the speed of its passage, and the word "place" refers to its position in the east, and what is said about "place" refers to the same point, unless it describes either a sphere or a spiral, which does not apply to the present case. That this is the genuine meaning of his words is clear from many others, including Olympiodorus in his *On Ecclesiastes 1*, who insinuates that what is said in Esdras was taken from Ecclesiastes, and is without doubt the meaning which he intended, as was said above in chapter 8.

Third Argument

The Scriptures speak metaphorically and in the language of common ordinary meaning in many places, but this is especially true in mathematical passages, where theorems and demonstrated truths are kept secret. The antecedent is obvious. For Genesis 1 [1:16] says that the sun and the moon are large lights in the heavens, as they are to the senses; yet they are smaller than the far off stars of the firmament. And Joshua 10 [10:12] says that the sun stood still in the middle of the heavens; yet it came to rest towards the west. And 3 Kings 7 [1 Kings 7:23] says that the diameter of the bronze sea from lip to lip is ten cubits and its circumference is thirty cubits; yet the diameter is not related as 10 to 30, but rather as 10 to 31. And 7 to 22. And likewise in other cases. Therefore, etc.

RESPONSE: I deny the consequence. The earth standing still and the sun moving do not have a metaphorical sense. For what would that sense be? Nor is that said in accordance to common custom, as is clear among other things from the solution of the above argument. Let me add that when Scripture speaks metaphorically, the literal meaning is not what the metaphor signifies, but what is symbolized. For when it speaks of God's eyes or arm, this does not signify that God has such parts, but rather it displays His power and operative force. See St. Thomas [*Summa theologiae*], part I, q. 1, art. 10. From this some say that the literal sense is not to be taken broadly according to what is written, or as what is signified by the surface of what is written, or even according to its logical or dialectical force, but rather according to the expressions, commonly used in rhetorical speeches or according to tropic or figurative expressions, which are commonly used when mixed together with a consideration of other nearby words, both before and after. See [Alonzo] Tostado, *On Matthew* 13, q. 28, and [John] Gerson, *Treatise on the Literal Sense,* proposition 2. Furthermore in regard to what pertains to metaphors, things that are mentioned in one passage in Scripture are clearly explained in other passages. See St. Thomas [*Summa theologiae*], part I, q. 1, art. 9.

However the Fathers also remained close to the mystical interpretations when they would propose a physical reason for the immobility of the earth or would explain it in the context of their own interpretation. Olympiodorus stands out as especially distinguished above the others when he explained the words of Ecclesiastes, "The earth stands still forever," as referring mystically to the immutable truth of Sacred Scripture. He says that Sacred Scripture is called the "earth" because it is an image of the unchangeable earth, it bears fruit, is made available to all, receives all in a friendly way, remains forever based on the old and the new testament, and contains immutable promises for the just and declares punishments for sinners. I will omit the rest.

The fact that the sun and the moon are called the great heavenly lights is explained by Chrysostom in his *Homily 6 on Genesis* as a reference to the magnitude of their power and efficacy over lower things. St. Thomas [*Summa theologiae*], part I, q. 70, art. 3, ad 4, also says the same thing about their magnitude as they are sensed, and speaks of the moon specifically because it appears so much larger. But there is no comparison of the heavenly lights with the stars, both because they are smaller than the sun and, because of their distance, their visual diameter cannot be easily seen, whatever frivolous things some have said about measuring them with a closed optical tube [telescope]. The moon is said to be large, either absolutely because of its conspicuousness and without a comparison to the sun, which could be larger, especially when it is on the horizon; or else the moon appears larger when the vapors in front of it are removed, although sometimes it seems smaller because of its high altitude which causes one to think that it stands still on a flat surface and the senses do not easily observe its unevenness or its sphericity. Meanwhile Vitellio [or

Vitello, *Perspectivae*], bk. 4, proposition 65, shows that because of its closeness the full moon appears to be swollen and puffed outward in the middle of its body, even though otherwise our vision always sees less than half of a hemisphere of the lunar body because there is a small separation between the eyes of the person measuring the lunar body. The same argument applies to the sun. Furthermore there is no deception of the senses in this case, for the heavenly lights truly are large to the senses, even though they are even larger according to reason. But if someone were to judge contrary to reason that the sun stands still and the earth moves, that would indeed be a true deception.

Furthermore, when it is assumed that the sun and the moon are smaller than the stars of the firmament, that is deduced from the immense distance between the firmament and the heavenly lights just mentioned. The defenders of the Copernican system imagine that, since the stars are seen at an almost infinite distance, they have a size which is hardly explicable by any proportion. Others say that it is true that the sun is even larger than any star of the first magnitude, and since the moon is smaller than all stars of the sixth rank, then it follows that the sun is larger than the earth. From this one could also account for the sun being the "larger light" and the moon the "smaller light" in terms of their relation to the stars of the firmament, although this is more properly understood as referring to the sun and the moon as compared to each other.

It is false to say that Joshua wanted the sun to stand still in the west, and not in the middle of the heavens. There is no sufficient reason to say this, as if Joshua feared that the enemy would escape under cover of night, and for that reason he restrained the sun from being an anticipation, or a forewarning, as the Hebrew text reads. For Joshua was able to foresee early that the day of victory could be lost, and he ordered the heavenly lights to stop in anticipation of that. Thus the word "stop" means that the sun stood still at or near the middle of the heavens, especially since the Septuagint text reads, "And the sun in the middle of the heavens did not go on to the west." This is clearly shown by reason. For it is said, "Sun, stand still above Gibeon, and moon, above Aijalon." From this it follows that both of the heavenly lights were above the horizon, but the moon was toward the west above Aijalon, which is west of Gibeon, above which the sun was ordered to stop. Thus the moon did not follow, but preceded, the sun by a distance of four hours or 60 degrees between them. The remaining 30 degrees were to the west, and the moon grew older gradually for about a 1/22 part of a day.

In regard to the diameter of the bronze sea [water tub], I say that it was made as it was, not because the Scripture was unable to provide the precise proportion of the circumference to the diameter [of a circle], but because it was sufficient for its purpose to request that a mechanical object be produced. Nevertheless it was written down beforehand that the sea should have no defect in its roundness, having ten physically proportionate cubits of diameter and an evenly curved circumference. And to be mathematically exact (as I said) in its roundness, it would contain equal sides and exact angles, and thus be most round. Moreover there was no mystical reason why it had to be just that size, and neither larger nor smaller. However if we inspect the implied meaning of Scripture more carefully, it is easy to establish the mathematical proportion more adequately.

For in the Hebrew, the Chaldean, and the Latin texts, 3 Kings 7 [1 Kings 7:23] reads as follows: "And a thin string of thirty cubits goes around it in a circle." But the Septuagint edition changes this to read "three and thirty cubits" and the emenders have changed the books accordingly. Take one palm from each side of the bronze sea. These two palms (by "palms" I mean "spans" or large palms) make one cubit, which, when added to the diameter, gives eleven cubits. Then divide each one into thirds, giving three and thirty cubits. Of these, thirty cubits in turn are cut in half, and gathered into six groups of small palms. The number which results, which is the number of small palms in the circumference of the sea, whose diameter has been measured with a golden bar, is a fourth number, namely, three and sixty. This is the number of palms which the sea has according to the Septuagint. Finally since they were distributed into six groups, the result is [6] tens and three palms. If from these the ten cubits of the diameter of the sea are removed, and so are the two palms belonging to the two sides, then one palm will remain which extends the lip. And from this is derived, by means of a more mathematical calculation, the ratio found in the Septuagint. See Theodoretus [Bishop of Cyrus] on 3 Kings 7 [*Patrologia Graeca,* vol. 80, 686–91].

[The calculation has proceeded in this way: 10 + 1 = 11; 11 x 3 = 33; (30 x 2) + 3 = 63; 63/20 = 3.15. The actual ratio of the circumference to the diameter of a circle, or pi, is 3.14159. . . .]

Fourth Argument

It is not contrary to Scripture to say that the firmament and the stars in it are at rest. Therefore a necessary consequence is that the motion of the earth is not contrary to Scripture. Or if the motion of the earth is contrary to Scripture, then the motion of the firmament, which many defend, will also be contrary to Scripture. For there are no more passages that say that the earth is at rest than say that the firmament is at rest. Furthermore the original Hebrew word for the earth is derived from a root which has the same meaning as "to run," clearly to denote the velocity of its motion, even thought it is imperceptible to the senses. Therefore, etc.

RESPONSE: I deny the antecedent if the word "stars" is understood to include the planets. For in this compound sense the proposition is false, and also is contrary to the faith, for a state of rest of the sun is contrary to the faith, as was shown in chapter 9. If the word "stars" does not include the planets, then whatever be the status of the antecedent, I deny the consequent. For if the motion of the earth is contrary to the Scripture, then the motion of the firmament will not also be contrary to Scripture, because there are many passages which clearly state that the earth is at rest, while it is not clearly stated whether the firmament moves or is at rest.

To repeat, it is false to say that the number of passages that assert the stability of the earth is not larger than those which assert the stability of the firmament. For we have established the opposite both from what was said above in chapters 3 and 6, and from the fact that the earth is the center of the universe, which was said in chapter 10 to be rather probably a matter of faith. At the same time it was shown that the earth is firmly at rest in relation to the planets, and also in relation to the other stars, even if the firmament does not move. Furthermore those passages that seem to insinuate that both the earth and the firmament are at rest do not give the same reason. For firmness is hardly ever mentioned without something being added to explain the type of firmness. Thus when it is said, "He firmed up the orbit of the earth which will not be moved," and "He founded the earth on its own stability and it will never incline," and many similar passages, the type of firmness is expressed in the words. Meanwhile the firmness of the heavens is simply stated, often without any addition being made, for example, only simply its creation is mentioned; or if something fuller, its order, location, or embellishment might be indicated. As a result the Latin interpreters have used many other words, for example, extension, expanse, and the like. Occasionally its embellishments are expressed in other words, as in Job 37 [37:18], "Have you perhaps helped him to make the heavens which are most solid like melted bronze?" Other editors read, "durable as a most solid mirror." This is not firmness in the sense of a state of immobility, but signifies with the greatest clarity a special solidity that is able to be crossed and that, though moved, is in no way slowed down; or more certainly and commonly, it signifies a liquid that in no way flows downward, nor do the planets and the stars in it vary from their regular paths to their preordained goals. On this point many have expressed great wonder and praise for the wisdom and power of the creator, who formed the heavenly body to be most thin and most fluid, so that it would remain unique and splendid and would be armed with a special stability against flowing downward. It is like melted bronze, which naturally exudes the splendor, smoothness, and cleanness of a metallic mirror, which the heavenly body is believed to be. Although some have thought that it was made from air, and others from a most subtle fire, others affirm that it was made from a special fifth essence distinct from the nature of the elements.

I have said before that whatever be the state of the antecedent, if the planets are included in the word "stars," etc. For even though nothing is said explicitly in the Scriptures about the specific motion of the firmament, still the whole proposition, "The firmament moves and so do the stars in it," is implicitly contained in Scripture because of its copulative structure. For God put the stars in the firmament, just as he put the heavenly lights there, to carry out their own role in their motion, just as the heavenly lights carry out their role of governing the days and the nights, as was said above in response to the First Argument. I submit that the firmament, when taken specifically, moves, and that this has a Catholic meaning, not only as the common opinion, but also because, from what was said above in chapter 10, it was the opinion of various Fathers, especially Dionysius the Areopagite.

Moreover, looking at this further, St. Chrysostom does not grant that the whole conjunctive statement, "The firmament stands still and the stars in it," is contrary to Scripture, even though he held that the heavens are permanently immobile. For in his *Homily 6 on Genesis* he says, "He put them (that is, the stars) in the firmament of the heavens, for what could 'put' mean other than someone saying that they were located far away? For we see that in one moment of time they cross through a huge space, and they never stand in one place but they follow the path that the Lord ordered for them." Thus says Chrysostom. And when the adversaries say, "The stars in the firmament always maintain the same order and relation to each other, but vary in their latitude in relation to the equator," what could they be trying to say other than that this happens because of the motion of the firmament? As a result, when combined, these two points, "The firmament is at rest and so are the stars located in it," cannot be verified, and neither can the contrary. Hence one of these two must be true: either the heavens move and the stars stand still, or on the contrary the stars move and the heavens stand still. For no one thinks that they both move at the same time.

What the Holy Fathers said about the motion of the stars is not sufficient to explain the planets. But both of these topics are explicitly and specifically discussed by Chrysostom, and I will remain silent about the others. He speaks about the stars in *Homily 6 on Genesis 1*, and about the sun (the same explanation applies to the other planets) in *Homily 12 to the People of Antioch*. He explains the motion of the stars, which we insinuated just above, both from the Scripture in Genesis 1 [1:17], "He put them in the firmament of the heavens to shine upon the earth," etc., which Chrysostom explains, and most clearly from the passage of Wisdom 13 [13: 2], "They thought that either fire or wind or fast air or rotating stars or mighty waters or the sun and the moon were God and the controllers of the orb of the earth." But the motion of the planets, which is not mentioned in either passage or elsewhere, can be best accounted for from the words preceding the prior quotation, "Let there be lights in the firmament of heaven, and let them divide day from night" [Genesis 1:14].

From this I infer the following. If the word "heavens" is understood to mean the firmament, the planets, and the remaining stars, then the two propositions, "The heavens exist and do not move," involve a contradiction, given the present order of things. Otherwise, if "heavens" is taken absolutely, either as including the empyrean heaven, or in the opinion of some as involving its own motion, without which its existence would be limited in nature, then it must move perpetually. Taken in that sense, the following propositions were once condemned in the Articles of Paris: "To say that the heavens exist and do not move is to utter a contradiction. For the existence of the heavens is simply to be itself, including the power to move itself. And the heavens preserve this power through their motion. Hence if they were to cease their motion, they would cease to exist. If the heavens were to stop, fire would not burn wood, because there is no God." From this it was inferred further that the generation of lower things, which is the purpose of the motion of the heavens, will never cease. And in the absence of a generator, it would be wrong to think that there is any motion of things that would then be at rest. Hence these propositions are also rendered false by the same passage. "Theologians who say that the earth is sometimes at rest argue from a false supposition. The heavens are never at rest. For the generation of lower things, which is the purpose of the heavens, ought not to cease. And the argument of the Philosopher [Aristotle] demonstrating the eternal motion of the heavens is not a sophistical argument." For it refers not so much to the denial of a purpose but to the beginning of motion. On each of these points see the Anathemas of Etienne [Tempier] of Paris, chap. 5, numbers 13, 17, 18, 25. [See *Chartularium Universitatis Parisiensis*, 1:543–55, for his 219 anathemas of the year 1277.]

Furthermore the claim that the earth does not move is in agreement with Scripture since its opposite is explicitly contrary to Scripture, as we have shown in chapters 3, 6, and 10. Therefore the consequent [of the Fourth Argument] can correctly be denied even though the antecedent be granted. The heavens being at rest is in a sense not contrary to Scripture because that is not stated in Scripture with undoubted openness, immediacy, without deduction, and in the same way that its opposite, the motion of the earth, is contrary to Scripture. Rather its contrary is found in Scripture in a partially open way, and is easily but partially derived from various passages, according to the interpretation of the Fathers. Some of them follow one philosopher, and others other philosophers, and as a result they have different views about the heavens and the earth. Some even think that the earth floats on water; but no one says that the earth moves in a circle. Nevertheless they all think the same thing, if not about the heavens as a whole, then at least about the planets and the stars, and especially about the sun, as is clear from we what was said in chapter 9. Moreover some things that the Holy Fathers have said repeatedly are less certain because of their different philosophies. "They do not treat some things which they have said in philosophy as if they are asserting them but as if they are using them. In such cases there is no authority greater than the sayings of the philosophers whom they follow." All this has been admirably said in the words of St. Thomas, *In 2 Sentences*, dist. 14, art. 2, ad 1.

In his *Commentary on De Caelo*, bk. 2, lect. 20, where St. Thomas examines the opinion of the motion of the earth, he does not say that it is contrary to Scripture, as he usually did on other topics, but at most that it is contrary to Aristotle. One reason among others that he did not evaluate it according to the principles of the faith, but only according to natural principles, was to make clear what he says elsewhere in a similar case, in *Opusculum 10*, art. 16, where he discusses the natural power of angels to move the mass of the earth up to the globe of the moon. For in such a situation St. Thomas's position was to neither affirm nor deny something, as pertaining to sacred doctrine, that the philosophers commonly think otherwise to be false. This is especially the case when there is no argument for the matter, and one could urge more time to find one, "lest otherwise an occasion be easily created for the wise men of this world to condemn the faith." When the matter does not intersect with the principles of the faith, it might be easy to refute it as absurd from natural principles. St. Thomas says this same thing admirably and frequently in his *Summa contra Gentiles*. See *Opusculum 10*, referred to above.

Furthermore this Fourth Argument can be easily refuted in these and other ways. The earth being at rest is not contrary to Scripture; indeed it is quite in conformity to Scripture; therefore the claim that the firmament and the stars in it are at rest, taken copulatively or disjunctively, is contrary to Scripture; hence to say that the firmament and the stars (taken as said) do not move is contrary to Scripture. These propositions have a certain mutual and necessary connection to each other, such that if the affirmative or the negative of one is granted, then the affirmative or negative of the other follows, and vice versa.

In regard to what was said about the root of the Hebrew word, I say that there is no interpretation, not even one, that favors what was proposed. Consider the following. The earth is figuratively said to run because, while it is immobile, the celestial sphere surrounds it and rotates around it, just as in Latin we say there is light because there is a little bit of light, or war when there is just a small war, or six hundred other things like that. Also the same root that signifies "running" also signifies "rubbing," which many claim is a certain type of fixed stability. Thus the earth is rubbed by all the feet walking on it (from which perhaps the Latin word was derived by chance). Or perhaps the earth was pushed, even though it remains immobile in the same place, or whatever else it was that was presumed by Archimedes, or whoever it was, who imagined that if a foot could be firmly planted outside the world, the whole orb could be moved in place. On the other hand the Greeks used the word "aether" to refer to the whole region of the stars which moves with an extremely fast speed. This view was held by the Greeks as one of the oldest truths of the senses. And for that reason the greater the ambiguity that can be found in the Hebrew root, the more one ought to accept the certain authority of sensible experience and of rational proof.

Fifth Argument

Hell is in the center of the earth, and in it fire torments the damned. Therefore it is totally necessary that the earth moves. The antecedent is obvious. For the Letter to the Ephesians 4 [4:9] says, "He descended to the lower parts of the earth." And Acts 2 [2:27] says, "Do not abandon my soul in hell," which clearly means the fire

of the damned. And Matthew 25 [25:41] says, "Depart from me, you who are ac-
cused of evil into the eternal fire which was prepared for the devil and his angels."
The consequent is proven by the fact that fire is a cause of motion. As a result
Pythagoras, who according to Aristotle put the place of punishment in the cen-
ter, thought that the earth is animated and is endowed with fire.

RESPONSE: I agree that hell is in the center of the earth and that it contains
fire. But I deny the consequent. To prove this I say that if it were valid, it would
also prove that over-heated furnaces, stones, ovens, bakeries, and stoves would
be animated and would move themselves. But I say that fire is not a cause of
natural motion unless it goes straight upwards by itself. And neither Pythagoras nor other more recent observers of the earth concede such a motion, although some try to allow for a circular motion of bodies around the center. If
we are speaking of fire in an animal body, this is a true cause of motion only
instrumentally and is due to the informing soul; yet it does not exist in the
animal formally, or in a weakened way, if we are talking about elemental fire
or fire in any other sense. If Pythagoras attributed such a soul to the earth because of the fire it contains, then he spoke out of ignorance and contrary to
his own principles, for elsewhere the stars are said to be fiery and animated,
and the sun is animated and has a fiery nature, and hence they would be endowed with motion. Then how can the sun be at rest in the center as the earth
moves around it? Or if the sun is said to move in a rotational motion on itself,
then the difficulties raised above in the solution of the Second Argument return again.

In his *Commentary on De Caelo*, bk. 2, lect. 18, St. Thomas shows with
many proofs that the earth is not animated, in opposition to the arguments of
Alexander who followed the errors of the Gentiles, who as a consequence attributed a cult of deity to the earth. Indeed every motion has some mover besides
its natural form. But in the motion of simple bodies, as is the earth, there is no
other essential mover besides its generator. Thus it is clear that the earth is not
animated, for its generator cannot be a soul joined to the earth, as Pythagoras thought. Rather its generator is a separated mover, which in the opinion of
Pythagoras and other ancients is not sufficient in the case of the earth, but requires furthermore a soul joined to the earth as a proximate mover. If the Pythagoreans wish to say that this is fire, they would also have to admit that this
fire is produced by a separated form and again the earth was produced by fire
according to this same philosophy. Everyone sees that all of this is ridiculous
and absurd.

Thus clearly the Pythagoreans have argued in many ways against their own principles, especially in what they think about the earth being built with forty-eight angles and containing six triangles with equal sides, and similarly with the cube. And as was said above in chapter 3, they give a completely useless account to the earth's motion; nevertheless they endow it with a soul and claim that it rotates with an extremely fast motion. This has led their followers into frivolous errors. Because of that they have been ridiculed by the philosopher Hermias in his *Satire on the Pagan Philosophers,* and especially because of their argument about the cube and the earth, they are not undeservedly listened to with laughter.

Some say, "Fire is contained by force in the center, and that by its own nature it seeks to be above the earth in a sphere as its own place, and that therefore the earth moves." I say that this has nothing to do with the way things are. For firstly such a motion would be accidental and not due to nature, which a law-like motion in the earth ought to be. Furthermore the fires of hell are capable of doing many things which elemental fire cannot do. Among other things they were established by a divine order of justice to be contained in that place forever, where their activity and preservation is maintained, especially by the coldness of the earth which surrounds them on all sides. See St. Thomas [*Commentary on 4 Sentences*], dist. 44, art. 2, q. 3, ad 2.

I have said above that hell is in the center of the earth. However in his *Dialogue [with Macrina on the Soul and the Resurrection]*, bk. 4, chap. 42, Gregory [of Nyssa] says, "I do not dare to speak rashly on this point." But nevertheless he thought that the opinion of those who say that it is under the earth is rather probable because, he says, "If we say that hell lies as far away from the heavens as the earth does, then it ought to be located at the earth." And also *Apocalypse* 5 [Revelation 5:3] says, "No one could open the scroll . . . and not those below the earth," which Gregory says "refers to the souls in hell." In *Opusculum 10,* art. 31, St. Thomas says, "In regard to the question of whether hell is at the center or near the surface of the earth, I think that this in no way pertains to the doctrine of the faith, and it is superfluous to be concerned about asserting or disproving such things." And in *Opusculum 11,* art. 24, he says on this matter that it seems to him that nothing should be asserted, especially since no one knows where hell is. He also says, "Nevertheless I do not think that it is at the center of the earth." But he would have taken back these words if there were some reason to do so, since he later added, "However it is not said that Christ descended to the lowest, but to the lower, parts of the earth, which would be true enough if those parts would somehow seem lower to us." And in article 25, given the supposition that hell is in the center or near the center, he thinks that it would be possible to know the distance from the surface of the earth to its center, but not the distance to hell, because he says, "I do not believe that humans can know where hell is."

Nevertheless I do not want anyone to infer that since hell is in the center of the earth, but we do not know where, then we also are ignorant of where the earth is, and consequently it cannot have been put in the center of the universe. Others will refute that argument. For if the location of the earth is unknown, it cannot be said to be in a large orbit moving around the sun. Also I say that as a matter of fact we do know where the earth is, both from all the different places determined by its fixed axes and from the observations and measurements of the geometers, even though we do not know everything contained in it and what specific and proper place each one has. For if hell is at or near the center of the earth, its general place is known, just as the location of the earth itself is known.

Nevertheless in his *De Genesi ad litteram,* bk. 2, chap. 34, St. Augustine placed hell under the earth, which passage he himself cited in his *Retractationes,* bk. 2, chap. 24. He says in book 12, "It sees to me to be better to say that hell ought to be below the earth, than it would be to give an argument why one should believe or say that as if it might not be so." And in his *Commentary on 2 Sentences,* dist. 6, art. 3, ad 4, while discussing the opinion of Pythagoras who claimed that the place of fire is a prison that he himself put in the middle of the universe, St. Thomas said, "We say that this is the location of hell." And in his *Commentary on 4 Sentences,* dist. 44, art. 3, ad 2, and in the "On the contrary," St. Thomas inferred that, "The fire of hell is below us," from the text of Isaiah 14 [14:9], "Hell down below was thrown into confusion by your arrival." And in dist. 48, q. 7, art. 3, ad 4, he said, "After the resurrection when the light of the moon will be truly increased, there will be no night anywhere upon the earth, except in the center of the earth where hell is located." And in dist. 50, q. 2, art. 3, ad 4, he said, "In the middle of the earth where hell is placed, there can be only confusion and torches of fire."

Moreover in *Opusculum 10,* art. 31, St. Thomas does not think that this specific point belongs to the doctrine of the faith. Yet Christ's descent into hell or the lower parts of the earth does belong to the doctrine of the faith; as also does the notion that the earth, in which hell is located, was put in the lowest place, and therefore in the middle of the universe according to what we have said above in many places, especially in chapters 7 and 10. It has been shown that it is rather probably a matter of faith that the earth is indeed the center of the world. It does not follow from this that God could not create many worlds and many earths, as some think is so because all the systems would have to incline toward the center of the universe. For in that case, each particular system would have to sustain itself according to all of its parts. No one part rather than another would incline towards the center, even if the parts are of the same kind, for two reasons: first, because there would be no major reason to favor one over another; second, because just as God established a law that the waters not surpass their boundaries, as Eustathius in his *Hexameron* says, "He ordered the aether and burning fire not to consume the whole orb of the earth," so also and much more so, if there were many worlds, a law would be set down for each one of them. In his [*Summa theologiae*,] part I, q. 47, art. 3, ad 3, St. Thomas says, "It is not possible for there to be another earth besides this one." But this was not said according to the power of logic or according to the active power of God, as Cajetan [Thomas de Vio] has noted. Rather it was said in accordance with the power put into created things in their present order, which is that divine wisdom has so ordained things that all matter is contained in this world, and there is no other cause of it, as Democritus thought when he said, "This world and an infinite number of other worlds were made by the collision of atoms." That this is St. Thomas's interpretation is clear from the body of the article. In the answer to the third objection he gives the cause of this impossibility, that is, "all matter is contained in this world." It is clear that he speaks here only about power, as explained by Cajetan. And when he adds, "Every earth would be moved naturally to this center, wherever it is," it is clear that he argues undoubtedly against Democritus, and not about another universe created by an ordaining wisdom. This opinion cannot be disproven by anyone. The thinking of St. Thomas on similar issues is sufficiently well established elsewhere. He does not deny that other species of angels could be created (whatever their status is as individuals), nor does he deny that there could be a very much more perfect creature, and an indefinitely more perfect creature.

Sixth Argument

In the Heavens there are waters above and below the firmament, Genesis 1 [1:6–8]. Hence the same is true of earth, since waters are held back only by the solidity of earth. Consequently the natural place of the earth is not the center, and it could be carried outside of the center by a circular motion in a large orbit.

RESPONSE: Firstly the consequent can be denied. It is sufficient if the waters are held back by the solidity of the firmament, and do not flow down from it. For that is the reason why it is called the firmament, that is, it provides an impenetrable boundary for the waters, and thus "firms" them. It would be much better if the waters were thickened, as if they were compressed into ice or crystal. Those who speak of a starry, icy, or crystalline heavens think that they are compressed waters or that these names refer to a kind of water; hence they call them aqueous. It would perhaps be more sufficient if the firmament were given the properties attributed to it by Eustathius's *Hexameron,* who thought that it was different from the first heaven, and was endowed with a more solid nature so that it could separate the immense and abundant force of the waters above and below it. This is said to be an accidental difference lest he seem to contradict what he said a little earlier. To signify how he saw it, Isaiah used the words, "The nature of the heavens is tenuous and neither solid nor thick." Elsewhere he says,"He made the heavens stand still like a vault," and "He firmed up the heavens as if they were smoke," thus distinguishing between wet and dry fluidity.

I have said that the consequent can be denied. For if we admit that the same things can be said about the earth which we said a little earlier about fire and water, it is not necessary to attribute to the earth all these same properties if the earth is located at the center. For if the earth were celestial, it would be free of all defects of matter, it would not include the generation of things, and it would specifically include the ordained purpose (if not also other purposes) of carrying on the system of the heavens. Those who deny that the heavens are a fifth type of essence could maintain this opinion. However those who attribute a different genus of matter to the heavens could maintain this only partially. For that type of matter could not be explained by the same specific forms as found in lower things, but rather by the highest genus that is applicable. Thus those who hold that the firmament has the nature of the four elements, although it is not composed of them, but is itself like a simple element, seem to many to make good sense. For they can speak of the waters above the heavens, and in the same way of the earth existing above the heavens or the firmament.

The teaching of St. Thomas [*Summa theologiae*], part I, q. 78 [68], art. 2, can be applied here. If the firmament has a nature like the four elements, then "waters" will not have the nature of elemental water, but is called "waters" only diaphanously. Similarly what can be said of an earth existing in that same place is that it is either of the same nature as our elementary earth, or it is of a different nature, and thus can be called "earth" only because its density and solidity are enough to hold back water. One might object that the weight of the earth would cause it to fall, rather than stand still there. But this would not happen both because the firmament is an impenetrable boundary and because, like the waters which are naturally heavy, the earth also could be said to be held above the heavens by divine power, as St. Thomas says in the same place in his answer to the second objection. But let us not quarrel over what God did miraculously rather than what the natures of things are able to do according to His laws, as St. Augustine warns us in his *De Genesi ad litteram,* bk. 2, chap. 1. In his *Opusculum 11*, art. 24, St. Thomas states that just as it must be said that the waters in the heavens are thin and not dense like they are around the earth, the same must be said about the earth itself. However according to the opinion of the Fathers, which we discussed above in chapter 9, there rather probably is no earth there. For no specific purpose can be assigned to it which the firmament itself does not accomplish. And one should not multiply systems that need many earths, when it is certain that the order of the present universe is so constituted that it forms a unity out of many parts, in which the lowest place is occupied by the earth or center located in the middle.

Furthermore, what we have presented here from St. Thomas agrees with what was received from God by the learned and illustrious virgin St. Hildegard. I have no fear of praising this woman as an author. Her writings have been approved and praised by [Pope] Eugenius III, St. Bernard, other writers, and other popes: Anastasius IV, Hadrian IV, and Alexander III. It is clear from their letters to her that, because of her sanctity and her knowledge of divine secrets, she was held in great esteem not only by the Emperors Conrad and Frederick, but by innumerable others, especially the Princes of the Church. Finally the whole Christian world looks up to her with good reason, and she is especially venerated in her native Germany. In book 38 of her *Quaestiones,* which was published by Wilbertus, Monk of Gembloux, she responded to question 2, which asked, "Since it has been written that God divided the waters which are below the firmament from the waters which are above it, must one believe that there are material waters above the firmament?" Her response was, "God divided the waters which were above the firmament in order to have the lower waters available to create the earth, and the higher waters available to create the upper regions. In the higher waters nothing increases or decreases, while in the lower waters where living things reside, there is increase and decrease, as in humans. But the higher waters remain in the original state in which God created them, and flow in their own circle and are material. But the lower waters are much more subtle and altogether invisible to our eyes, because of the humidity and the fire that exists there. The firmament is solidified from above, like a body that exists because of a soul and does not dissolve. The lower waters under the firmament are heavier, and the celestial lights, namely, the sun, moon, and stars, are visible because they contain lower animals of different species, which are born and live their lives there. Therefore the nature of the higher and lower waters are, on the whole, dissimilar."

So says Hildegard. Her remarks include a circular motion of the heavens, water existing above the firmament without any overflow to the lower waters, and an insinuation of their thinness; and thus the waters differ in this way among others from the waters which remain around the earth. She says that the upper firmament is solid because of the humidity of the waters and the heat of the fire there. She also says that the heavens are liquid, according to the old and more recent common opinion, and moreover are beyond the senses of the viewer, as we mentioned a little earlier from Eustathius. "He firmed up the heavens like steam" is rather clearly explained by comparing its firmness to smoke and by assigning its cause to humidity and the heat of fire. We more freely affirm this view and its authors, namely, Eustathius of Antioch, Isaac of Syria, Alexander Lycopolis, and Hildegard, than the view of Christopher Scheiner in his *Rosa ursina*, with which they partly agree and partly do not. Nevertheless in his book there are many things which are defended here, and which can be pursued there in further detail.

Furthermore Averroes has presented a similar type of argument against eccentrics and epicycles. "If eccentrics are granted, each one will move around its own center, but an earth lies at rest in every center around which the heavens move, for everything that moves needs something that lies at rest." But this argument is philosophically weak. For if we grant that eccentrics have the same center in relation to their whole orbit, and if for epicycles it is sufficient if the middle point is taken, then the sun and perhaps also the other planets, which revolve on themselves, still all move around the earth as a whole. Granting all this, it is false to say that an earth must lie at rest in every center around which the celestial orbs move. For the earth could be carried away from the middle, yet the heavens still move around the center of the world, just as both eccentrics and epicycles move in their own special way around the middle which corresponds to them, whatever that center will be like in the future, either some other body, or most likely rarified water or air or an imaginary point in empty space.

The words, "Everything that moves needs something that lies at rest," are to be specifically understood in regard to bodies, insofar as they either move efficiently on their own or they presuppose a motion without which they cannot act. In this sense Claudius Mamertus said in his *De statu animae*, bk. 1, chap. 18, that nothing can move unless there is something that is unmoved. Thus a finger does not move unless the hand lies still, nor an arm unless the shoulder stands still, nor in walking can one foot move unless the other remains fixed. But in circular motion it is not necessary that something at rest accompanies the motion efficiently, but only that the motion occurs around a middle on which it depends, and thus stays in the middle and not outside of it. The middle is to be compared in a way to the stationary foot, and as the lowest part of the universe, on which the motion of the other parts depends. Hence if any other circular motion requires a middle at rest, then certainly the revolution of the heavens around the earth requires it.

There is also a very similar argument in Copernicus's *De revolutionibus,* bk. 1, chap. 10. The earth cannot be the center, or be in the center, in respect to the universe. For if the universe is infinite, it will have no middle point from which equal lines could be drawn to the periphery. If it is finite, the middle point will be indivisible and imaginary, which is not the earth since it is enormous and vast. If this language is related to the parts of the universe, the planets first of all are the eccentrics, for they do not have the earth as a center, but each has its own center. And the other parts of the universe also have their own centers toward which alone they incline, and not toward the center of the universe.

But this is a weak argument because it proceeds from the properties of the finite to the infinite (which is a sophism, as we have shown elsewhere). It is as if one were to argue from the human body to an angelic substance, or to conclude from a spherical and circular figure that the world is infinite because it has no beginning or end. In his *Liber de mundi creatione contra philosophos,* Zacharius [Rhetor, or Scholasticus, Bishop of] Mitylene has brilliantly ridiculed doing philosophy with such geometrical sophistries. If the truth be conceded, the universe indeed is finite and spherical, but it does not necessarily follow that its center is imaginary and indivisible. The latter speaks of a mathematical middle and of bodies in abstraction from motion and place. But the real middle is physical and in a place, and in the constitution of the material universe it is a physical body having three dimensions. Although it is very small in comparison with the celestial orbs, nevertheless it is such that the motion of the whole universe somehow depends on it. Even if it is conceived as a mathematical point, it possesses in itself immobility at that spot, not because mathematics has abstracted it from motion, but because what is indivisible cannot move except accidentally. And I say once and for all that arguments that are applied to physics from mathematics often lead the careless into sophistry. For example, this is the same way that Theodosius [of Tripoli] in his *Sphaericae [doctrinae propositiones]*, bk. 1, proposition 3, and his followers show that, since a sphere at rest touches a plane at only one point, it is possible to demonstrate that the same thing happens in motion, and consequently indivisibles gathered together compose a quantity. So it is amazing that the Copernicans have not noticed that almost the same argument in every respect can be made if we place the sun at the center of the universe.

In regard to the parts of the universe, each body considered individually has its own center to which the parts incline, just as water forms spherical drops. And each element is moved to its own place by nature. Yet since it is also part of the universe, it has a common center with the universe, which is the earth. And as I have said, it is integrated with its own parts for the reason that the universe itself be integrated and agree harmoniously regarding its site and order and other things of that type. Thus the planets do not move in eccentrics or epicycles unless, as we said above in chapter 10 regarding the whole orb, they have the same common center, that is, the earth. The earth is so related to the elements that it is heavier in weight than all the others and, strictly speaking, it underlies all of them. No one can easily deny this. For it is evident that bodies are spherically arranged and rotate in circles one after the other, that the heaviest one must be in the middle of the others, and that the universe has been constituted with this arrangement. Contrary to the opinion of the recent Pythagoreans, I submit that the most fit of all the elements to be at rest in the middle is the earth. It is completely lacking in heat, whose cause is motion, and therefore is an extremely cold element. It is not enough to say that outside its own place the earth would fall downward in a straight line, but in its own place it moves circularly around its own center, which is its natural motion, and otherwise it rests peacefully opposed to straight line motion downward. For the earth is a simple body performing one operation with the most simple motion. It has not been given many motions from its source for the whole is not generated. There is nothing to prevent it from returning if it happened to be held away from its own place. It has this property from itself, and hence there must be a distinction between its moving part and its moved parts. The former moves the earth in one way rather than another, while the many other parts mutually oppose each other, as the Copernicans agree in regard to the daily and the annual motions. How this can happen, or if the earth is to be impiously said to be endowed with a rational soul, no one can ever understand.

What some have seen in the words of Wisdom 19 [19:18], "The elements turn into each other in the way that the sound of a rhythm changes in a musical instrument," is false. For the words in this passage are not about circular rotation, but as the Greek text makes clear, they are about the adaptation of the elements to each other. In harmonic transposition, given that the sounds are mutually arranged to blend together, nevertheless by transmutation and conversion in a psaltery or any musical instrument, the sounds change their concord and rhythm but still retain some harmony. Likewise the elements prreserve their site and order, but nevertheless for some reason they are transmuted into each other. Or if this passage is to be interpreted to be truly about sounds, as some seem to have concluded from various readings of the Greek text, then it does not refer to the earth as moved, but to the higher aether. In her *Quaestiones*, bk. 38, q. 27, the illustrious virgin St. Hildegard has very clearly said that these words, when joined to what we quoted above in chapter 4, must be understood to be about sounds metaphorically, that is, about the concord of the elements as a harmony in relation to their order, site, and the other things that they have in their constitution.

SEVENTH ARGUMENT

The opinion that the earth moves and that the sun is at rest at the center is very old. It did not originate with Pythagoras but with Moses, from whom it was derived by Pythagoras, who was a Jew. This view was accepted not only by very learned pagans, but also by Christians, including Cardinal Cusa and others before Copernicus. Afterward Copernicus so restored this opinion that he can almost be taken to be its author. His book was approved by Pope Paul III, to whom it was dedicated, and was ordered to be printed as a great value for Catholics. From the very exact calculations of Copernicus, who used the Pythagorean system, both the Gregorian calendar and the very exact celestial tables used everywhere today were constructed. The assumptions of the motion of the earth and the stability of the sun must be true, because evident truths are deduced from them, and truth follows only from what is true. Many outstanding persons rightly turn to Copernicanism daily, and it would be difficult to convict all of them of error or boldness. Therefore, etc.

RESPONSE: Opinions are not proven because of their age but because of their truth; nor are they condemned because of their novelty but because of their falsity. Old is the opinion of Plato that the angels are material; old is the opinion of Anaxagoras that the sun is a burning stone; old is the view of those who say that the sun is a prison for the souls of sinners who are shut up in the body of the sun. Finally, lest I go on forever, old is the opinion of the Manichaeans who said that the sun and the moon are ships assigned to cleanse the world of its filth, as we find in St. Augustine's *Letter 119 [55] to Ianuarius,* in his *De haeresibus ad quod vult Deum,* and in his *De civitate Dei,* bk. 11, chap. 24, and bk. 20, chap. 16. Yet all of these opinions, and innumerable others like them, are false and disproven.

The claim that Moses originated this opinion is an imposture. This is like one of those empty and foolish inventions, or views often used for nefarious purposes, which many ascribe to Solomon and to the words of Sacred Scripture. Although there is no heresy here, this does not protect Sacred Scripture from being poorly understood and distorted. Further, if Moses were the author of this doctrine, which is clearly most false and contrary to Scripture, then the same could be said of what he wrote in Genesis 1, as we have shown above in various places. And does it make any difference whether Pythagoras was a Jew or a Gentile, or whether Marcellina with Carpocrates burned incense to him? Also it is well known that few of the ancient philosophers read or understood Moses, and they corrupted many things when they translated his narratives.

I admit that there are many followers of this opinion, and especially of Copernicus. But there are in contrast many more who embrace the common system rather than the Copernican system, when you compare them in nature and in number. This larger group indeed works carefully and prudently to develop astronomy, and they are to be conscientiously imitated. They do not construct their system as naturally certain, and when they recognize that it is not in agreement with Sacred Scripture and the Holy Fathers, they take it to be only hypothetical and proceed with their calculations, knowing that that is all they are permitted to do. We think that in the long run their work will have better results if they maintain the truth and take another road and line of thought, and if in their work to develop mathematics and to serve the Christian community, they carefully consider how to avoid conflicts with the Scriptures. (As the author of the Frisian tables says on page 318), "The Sacred Scriptures should have such great authority among us, and our minds should be moved with such great reverence for it, that we would not dare to fall into the opinion of the Pythagoreans, which is openly contrary to Scripture."

It is reckless to say that Copernicus's opinion was approved by Pope Paul III. I do not deny that Copernicus was acknowledged by many for his talents, and that his work with calculations was praised for being of great value for mathematicians and for determining the calendar. He would have been praised even more if he had presented his work without an unconditional assertion of his system and theory, which he could have done. Although his work was accepted by the pope for the reason indicated when it was first published, later other popes repeatedly either ordered or permitted it to be suppressed, and it was never believed or affirmed that it contained nothing contrary to Catholic views. Indeed to avoid others from being led into error, the Sacred Congregation of the Index has not too recently issued the following warning: "The Holy Fathers of the Sacred Congregation of the Index have decided that the writings of the noted astronomer Nicholas Copernicus on the revolution of the world are to be entirely prohibited, because his principles concerning the location and motion of the terrestrial globe are contrary to the true and Catholic interpretation of the Sacred Scriptures (which is completely forbidden for any Christian), and because he does not hesitate to present this as quite true and not merely as an hypothesis. Nevertheless, since many things very useful for the human community are contained in them, the Fathers have by unanimous consent arrived at the judgment that the writings of Copernicus published up to the present are to be allowed, as they have been allowed, as long as the passages, in which he discusses the location and motion of the earth affirmatively and not just hypothetically, are corrected by the changes listed below. Furthermore editions published in the future must be changed in the indicated passages as follows, and these corrections are to be added to Copernicus's preface." So said the Sacred Congregation [15 May 1620].

In constructing the Gregorian calendar Copernicus's calculations were not used to the exclusion of older calculations. Aloysius Lilius was the first to propose the best way to reform the calendar. But he used the Alphonsine tables to define the length of the year, which was an average and hence more certain or less exposed to errors. From this he arranged the intercalations so that the equinoxes would always have a fixed and certain location. Since that is desired for human affairs, others used the Copernican calculations. But the best practice is that today's formulators of ephemerides accommodate themselves to use every method of calculation, since that seems to correspond best with reality. Hence one can no more infer that the Copernican theory is true than that other hypotheses denying it are true, when the latter give the same result. For these are merely different ways of deducing the same conclusion from different syllogisms. Indeed when Clavius completed and finally published the Gregorian calendar, just one of his accomplishments in life, he used the Copernican system in such a way as to never once bring about a conflict with Sacred Scripture, and he openly professed that he did not acknowledge the different motions of the earth that its authors imagined.

Those who have constructed ephemerides did not hesitate to use Copernicus's calculations, since he, being aware of the errors of others, presented his tables as more correct, but not absolute. His tables were inadequate to establish complete ephemerides, and they were deficient in many ways that afterwards more recent scholars have tried to correct, beginning with Tycho Brahe. Indeed they did not make any special effort to use Copernicus's calculations or his system, but stayed with Ptolemy's fundamentals, including an immobile earth, and they either modified the true system or made their calculations in new ways when they encountered the new theories. Without doubt Tycho's hypotheses about the moon, derived partly from Ptolemy and partly from Copernicus, were so expanded with new additions that his discovery would clearly have been remarkable and more certain and more useful in the future, if he had lived longer. More recent mathematicians do not use Copernicus's calculations to the exclusion of Tycho's. Clearly in our era Tycho's tables have been used to calculate most exactly the motions of the sun and the moon, and the times of all syzygies and eclipses. As a result many have used these same results to establish with ease lunar and solar tables, and to determine with a simple method the Ptolemaic equivalences, once thought to be anomalies, even though there is a difference between the added or subtracted times and their exactly calculated times. Copernicus's tables are more laborious to use because of the double anomaly of the apogee or eccentricity and the average motion, to say nothing of the many added and subtracted motions of the sun.

If we look at other topics, the arguments of Copernicus will not seem to be preferable to the calculations of others which are of equal value. For example, in regard to determining the true syzygy, Ptolemy, Alphonsus, and Copernicus are nearly the same, but Tycho is better. In regard to the motion of the sun Alphonsus, Copernicus, and Tycho do not differ very much, but they disagree in their determinations of the length of the solar year. Also they agree on the interval between equinoxes, on the additions and subtractions made to them, and occasionally on the average motion; but they disagree on the reasons since they locate the motion closer to the heavens than does Ptolemy, while Tycho's work on the heavens is considered to agree entirely. Consequently they are wrong who say that present day mathematicians should use the calculations and arguments of Copernicus rather than those of more reliable authors. As examples, I point to Kepler and Magini, whose calculations are thought to have provided something more certain.

In regard to the maxim mentioned last above that "truth follows only from truth," it is a fallacy to apply this specifically to Copernicus's hypothesis. For the causes of the appearances are accounted for from the position of eccentrics and epicycles no less than if one imagines that the earth moves and the sun is fixed at the center, as has been well understood now for many centuries. Today we should not regard eclipses and other phases and apparitions to be any more certainly understood than Alphonsus, Ptolemy, and other prominent experts did. From the positions of the appearances some numbers can be deduced, which in our age do not correspond to the motions or to the positions themselves, but a satisfactory understanding of the motion of the celestial bodies cannot ever be determined. For who has ever correctly deduced the precession of the equinoxes? And what mortal has determined without error the period of one planet, so that as a result many tables could be constructed for each century, which would be certain for both public use and for the mathematicians? The words of Ecclesiastes 3 [3:11] are indeed true, "He gave the world over to their disputes." Now if the Copernicans can determine the positions in the future, this can be done more certainly by Ptolemy, Alphonsus, and others, who through the use of eccentrics and epicycles have saved, and will save in the future, all the appearances for many ages. And for this reason there is no need to invent any other theories that upset the system of the universe, especially when they agree neither with true philosophy nor with the Sacred Scriptures.

Further if the maxim "Truth follows only from truth" is attributed to Copernicus's theory as a unique cause in a true demonstration, this is likewise a fallacy. For the same true conclusion follows from other premises, which are easily shown to imply nothing that cannot be explained. On the other hand Copernicus's hypothesis involves many serious difficulties, although we will not expand on its conflicts with Scripture, which will be shown in another place. Meanwhile we note here that the physical causes of the appearances are not the various positions, but the various motions and periods of the stars, which are determined by many observations and experiences of these positions, from which the fixed ratio of the motions is established. As a result various syllogisms, with their various arrangements of terms and premises, are not the cause of various events occurring in the things under discussion. Rather the cause is the nature of these various things, whose properties come to be known by the corresponding arrangement of the terms and premises from which the conclusion is drawn, showing that its predicate agrees with its subject. Thus it happens that the same thing is concluded from different syllogisms, which are taken to be certain even though in reality the suppositions of the terms are false.

For example, someone might think that it is certain that all stones have sensation, and that all humans are stones; or again that all plants have sensation, and that all humans are plants. From both arrangements of terms a true conclusion follows in exactly the same way, namely, that all humans have sensation. Nevertheless the principles from which this is deduced are false. The same is true of our topic. The systems of both Ptolemy and Copernicus imply that observed events occur in various places and times, and yet each system is false and imaginary in that many daily positions must be invented. Ptolemy gave the same type of explanation of solar phenomena both with and without epicycles. Many explain certain apparent properties of the stars by using the motion of trepidation, while others use retardation. Likewise cosmographers use different names and assign different positions for longitudes and latitudes, while equally certain conclusions are drawn by using their commonly understood meanings. The same thing would happen if one were to imagine that the center is located at the pole and that the larger circles are arranged differently; or if one were to assign in various imaginary ways different places for fixed points of rotation. Nevertheless all these things are false, even though they would be verified if reason presupposes that positions are accommodated accordingly. The above arguments show how it is certain in logic that, from an understanding of the subject matter and of the consequent, only truth and never falsity follows from the true, while from the consequence and the arrangements of the terms, both truth and falsity follow from the false.

These remarks about arguments hardly need to be examined here. When these issues are set forth, many defenders of the opposite view then quickly take refuge in affirming that the motion of the earth and the stability of the sun is true (as they say) "according to philosophy, whatever it be according to theology." Good God, what kind of philosophy is this? Using this clever distinction, they think that the whole matter has been settled, not realizing that nothing is true in philosophy unless it is also true in theology. "Truth does not contradict truth," as was said in the decree of the [Fifth] Lateran Council, session 8, since all created truth is a participation in eternal truth, which contains no impurity, no falsity, and "where there are no obscurities." In the same place it is rightly said that this distinction is "philosophically rash," and in a word is a "plague," which would be justly said to have been born from ignorance. As a result that same decree issues a warning, to those whose business is sacred matters and to those who are committed to them by oath, that they should not undertake "the study of philosophy or poetry without some study of theology or canon law, with which they will be able to purge and heal the imperfections of philosophy and the foundations of poetry." Since this warning must legally apply to all, then mathematics should either not be dealt with at all, or only by those who are expert in true philosophy and theology, lest fallacies arise from ignorance of these matters and one falls into many errors. In general it is better to abstain than to deceive or to be deceived.

Finally it should be noted that there are those who loudly but falsely complain that every path of thought for the development of mathematics is closed if the Copernican system cannot be pursued. Yet before Copernicus had appeared on the scene, the philosophy of the Pythagoreans had been examined by many, but it was not introduced into the schools because it was opposed either to nature or to the divine faith. However in every age there were scholars who cast light upon the stars by their glorious labors, even if they did not illuminate all the difficulties. But Copernicus did not consult them so that he could avoid falling sometimes into darkness. We will be permitted to apply our talents to new theories to determine how closely they come to the truth, if they fail to reach the full integrity or science, but not to overlook a theory's terms, which causes one to lose one's way into absurdities and conceptions that do not agree with Christian piety.

In general philosophy reaches its greatest level of learning when its clear arguments allow it to see famous errors, for example, that there are three distinct substantial souls in each person; or that there is one rational soul which informs all humans and that it is mortal in each of them; or that angels are corporeal; or that there are other worlds existing in the sun and the moon, and also other creatures endowed with reason and humanity; or countless other loudly proclaimed monstrosities which might be introduced into the schools and be defended with obstinate speculations. Yet, leaving the faith aside, although many such things are disputed in philosophy and have some similarity to the truth, they cannot be proven by even one argument. Moreover there are enormous difficulties in every one of the disciplines, which would call every outstanding mind to battle for the public good and the glory of the creator. But it will not be necessary to pledge your allegiance to this or that teacher, if you do not deviate from the Sacred Scripture and from its true and Catholic interpretation. Nor perhaps should anyone be fearful and loudly complain that he is forced by the whip to accept the opinion of Aristotle and St. Thomas. But to him I would say with Tatian (*Oratio contra gentiles*) that, "the old wives' tales of Phaerecides were followed by the dogmas inherited from the Pythagoreans, followed by their imitation by Plato." But he wished to be committed to no one, to be located far from Aristotle, to be a wandering Pyrrho, and not to be a philosopher.

We think that this brief treatise is satisfactory for our theological purposes; others will argue mathematically in favor of the traditional system, leaving to us however part of our duty to be completed later. For now our attention has been focused on what St. Thomas said at the beginning of his *Opusculum 10.* "In regard to matters which philosophers commonly accept and which are not contrary to our faith, the safer course is neither to assert them as dogmas of the faith nor to deny them as contrary to the faith, lest the wise men of this world be given an occasion to condemn a doctrine of the faith." In some cases we have granted that the whole matter should be judged by wise men, and in other cases we have said that it should be referred to religion. But we have shown that the faith of the Scriptures is opposed to the view that the earth is in motion and that the sun is at rest in the center, a view commonly held by some philosophers, both Pythagoreans and Christians. This is especially the case for those who can be included in the grave warning given by St. Augustine when he speaks about Faustus the Manichaean in his *Confessions,* bk. 5, chap. 5. "What he has said about the heavens and the stars and about the motions of the sun and the moon is known to be false. And although these matters do not pertain to the doctrine of religion, still it is quite clear that his boldness is sacrilegious. For he not only is ignorant of these matters, but is also wrong, and he proclaims these falsehoods out of an insane vanity of pride, endeavoring to attribute them to himself as a divine person." For indeed there are those who are deceived when they examine the Copernican system by its many unusual errors. And it happens that they take pride in this, not as the result of human wisdom, but as though many things have been divinely revealed to them.

As a result when we undertook to write this treatise defending the stability of the earth and the motion of the sun, which pertains in this age to the teaching of piety and the faith, it was our intention to think piously and as a Catholic, and to try to write and to affirm this openly. In general one should take up the pen in obedience to religion. However in these matters we adjust our views to the judgments of the wise, and we submit them especially to the evaluations of the Holy Mother Roman Church, for we do not wish to say or to assert anything which the Church thinks should not be said or asserted. Let us recall the advice of St. Augustine in his *De Genesi ad litteram,* bk. 2[1], chap. 18: "We should not go into battle for our own opinion rather than for the views of the divine Scriptures, wishing that they would conform to our opinion; rather we ought to wish that our opinion corresponds to that of the Scriptures."

Praise be to God

Jesuit Rules on Theology and Philosophy

A: Fifth General Congregation of the Society of Jesus
(3 November 1593–18 January 1594)
Decree 41

After the deputies assigned to review educational policy had carefully discussed the matter for many days, they presented to the Congregation their report on the adoption of opinions in theoretical matters, which the Congregation approved. First by unanimous consent of all the deputies, it was decreed that our professors must follow the doctrine of St. Thomas as more solid, more secure, and especially as that was approved by, and is in agreement with, our Constitutions. Second the Congregation decreed that our professors must exactly observe the rules which have been formulated by those same deputies from the *Ratio studiorum* to govern the adoption of opinions in both theology and philosophy. These rules are as follows.

RULES FOR THEOLOGIANS ON THE ADOPTION OF OPINIONS

1. Our teachers in scholastic theology are to follow the doctrine of St. Thomas. Only those who are well disposed toward St. Thomas are to be promoted to

chairs of theology. Those who poorly understand him, or who are not in sympathy with him, are to be relieved of the office of teaching. On the questions of the conception of Blessed Mary and the solemnity of vows, they are to follow the opinion that at this time is more common and more widely received among theologians.

2. The primary concern in teaching should be to strengthen the faith and to nourish piety. As a result, on issues not explicitly treated by St. Thomas, no one should teach anything that is not fully in agreement with the intentions of the Church or that would in any way diminish the strength of solid piety. Relative to this, our teachers should reject arguments that are not yet accepted, even though these agree with the arguments customarily used to prove matters of faith, and they should not casually introduce any new ideas unless they are based on fixed and solid principles.

3. Further, even in matters that present no danger to faith and piety, no one should bring up new questions of any importance, or any opinion not identified with some author, when inexperienced people are present. They should not teach anything that is contrary to the axioms of theology or to the common understanding in the schools. Rather they should follow the most universally recognized teachers and the most widely accepted teachings in the Catholic schools, as shown by the passage of time.

4. No one should teach or defend those opinions by any author that are known to be gravely offensive in some province or in Catholic schools. For although the teachings of the faith and the integrity of morals do not lead to division, prudent charity demands that we accommodate ourselves to those who are versed in such matters.

5. If Catholic Doctors do not agree with each other in a case where St. Thomas's opinion is ambiguous or where he perchance does not treat a topic, our teachers may follow either side of the question. However they should defend both sides of the issue equally, and they should weigh with special care, honesty, and good will the view of an earlier professor who has taught the opposite. Also indeed if the authors can be reconciled, it is quite desirable that this not be neglected.

RULES FOR PHILOSOPHERS ON THE ADOPTION OF OPINIONS

1. In matters of any importance professors of philosophy should not deviate from the views of Aristotle, unless his view happens to be contrary to a teach-

ing that is accepted everywhere in the schools, or especially if his opinion is contrary to the orthodox faith. In accordance with the [Fifth] Lateran Council, they should strenuously try to refute any arguments of Aristotle or of any other philosopher that are opposed to the faith.

2. Unless they have a very good reason, philosophy professors should not publish or present in the schools interpreters of Aristotle who find fault with the Christian religion; and they should take care that their audience is not influenced too much by such authors. For the same reason digressions on Averroes (and other authors like him) should be put together in a separate treatise. And if anything of value is found in him, they should present it without praising him, and if possible, they should prove the point in some other way.

3. They should not commit themselves or their students to any philosophical sect, like the Averroists, the Alexandrians, etc. Nor should they conceal the errors of Averroes and others; rather they should sharply suppress the authority of such authors. On the other hand they should always speak of St. Thomas with great honor, which means following him willingly in thought as often as possible, and departing from him reverently and unwillingly when one disagrees.

4. Philosophy professors should not introduce any new questions, or any opinion which is not attributable to some author, to those present who are inexperienced. And they should not defend anything that is contrary to the axioms of the philosophers and the common understanding of scholars. Those who are prone to novelties, or who are too free in spirit, should understand that they will without doubt be removed from the office of teaching.

5. In regard to questions in which one is free to follow either side, both sides should be defended, and especially one should consider with care and good will the view of a previous professor who taught otherwise. And when it is possible to reconcile the two views, this must not be neglected.

B: Letter on the Solidity and Uniformity of Doctrine, from Claudius Aquaviva, S.J., to the Provincials, 24 May 1611

No great effort is required to realize that solid and uniform teachings are necessary for the well-being of the Holy Church, for the commitments and reputation of the Society in regards to its ministries and related enterprises, and for the unity and fraternal good will of all the members of the Society. As a result, from the very beginning of my administration of the Society, I have made

a special effort to follow the sense of our *Constitutions* so that it would be observed as closely as possible, as Your Reverend Father knows, and to study our
book *Ratio studiorum*. Moreover the better providence of God has provided a
more effective remedy in the decree [41] established at the Fifth Congregation in regard to following the teaching of St. Thomas. It was hoped that this
would have been sufficient for the intended goal. However, experience clearly
shows us that, despite the diligence that has been exercised to date, the Society has not attained what was hoped. This is partly due to the sheer number
of our writings, in which it is not easy to find uniformity, especially if one con-
sults those written before the Decree, and partly due to the wide variety of
thinking and of defended opinions that are taken as probable. At any rate, since
I truly believe that the Society is threatened much more now than ever before
by great dangers from this situation, I wish accordingly to establish a fully effective remedy. Our holy founder, full of the divine Spirit, anticipated this,
and with the strongest words provided protection in our *Constitutions,* part 3,
chap. 1, no. 18.

Indeed we have discovered that some of our teachers have freely embraced
and published a doctrine simply because it has not been declared erroneous
by the censors. And they think that they can defend it by their own wits; even
the censors, thinking the same way, have let many things pass, as if "solid and
uniform doctrine" means "can be defended as free from error." And there are
even those who think that, as long as we agree among ourselves on doctrines
and conclusions, we can use a great variety of means to support them, and they
devote their talents to this. But it must be carefully noted and realized that,
even though whatever is used must be true, it is not permitted to contrive new
ideas to defend a conclusion or to use a method of defense that entails new
principles. This is a danger to be avoided. I could give many examples to make
it clear that, in regard to conclusions, to principles, and to the method of defense, dangers arise from variety, from novelty, and from doctrines that are less
than solid.

Let me add that it is one thing to depart from St. Thomas in the case of
some conclusion that is solidly connected with the principles and authority of
the most important ancient writers, which seems to be permitted by decree 41
of the Fifth Congregation, and quite another thing not to follow him in other
cases where other more recent authorities are taken as foundational. As a result at the end of disputations often neither a Thomistic nor a uniform teaching
has made its appearance. The product is of such varied colors that no one who
knows the teaching of St. Thomas would judge that it has one unified shape.

Nor is it satisfactory to defend a position with two or three quotations taken from various passages and arranged in a sequence that is either inconsistent or forced, as though we are to believe that this is the opinion of St. Thomas because the author somehow indicated that it is contained in those passages. Rather it is necessary to see what St. Thomas thinks in the place where he explicitly treats the topic, and to examine carefully what he asserts, either in agreement or in disagreement, in the rest of his writings. Indeed it would seem to be especially difficult to establish that a doctrine is Thomistic when it is judged by scholars versed in his teachings to be new and foreign to him, and where a great deal of subtle effort is needed for one who studies the matter to determine what is genuine and distinctive of him in the words used.

Therefore, for the honor of God, we must be very careful and vigilant, and the censors must be more conscientious. If we do not contain ourselves within definite and accepted limits, we will face new threats and dangers daily. There is no doubt that we dishonor ourselves when someone wishes to proceed further when he sees that there is permission to hold something about matters that require many distinctions and great acumen, and then others do the same thing, with the result that nothing is ever seen to be stable and uniform afterwards.

Thus since it is obvious that a topic of such importance must be carefully considered and evaluated, I see myself obligated by my office to call for a new and more effective diligence on this matter. For that reason, in addition to what has been said above about the censors, most of whom we have appointed, I hope that in each province our teachers and other responsible members will zealously accept this with the same concern which they have for the honor of God and for the good of our religion. I also hope that, putting all reservations aside, they will focus their attention, not on what one or other of our writers has already published, but on how this true and real remedy can be applied wherever needed. Therefore, Reverend Father, I ask you to summon in your province six or eight of your more responsible and more learned Fathers, and to obligate their consciences with the contents of this letter so that they will think seriously about our remedy. After this, and in consideration of the many universities, schools, and lectures that our Society sponsors, let the Provincials say and write whatever occurs to them regarding the good or bad effects of this policy on the Church as a whole. As I have said, it is not sufficient simply to avoid error. We have not said this. And if perchance someone does go astray, which I believe is unlikely, the ordinary remedies are available. But this warning is necessary and helpful for the future.

After giving this matter your careful attention, Reverend Father, send us through the Procurator your views and those of the above mentioned Fathers, so that we may decide what is most useful for the Society. And may this abound to the greater glory of God, for whose rich graces I pray for all of you.

C: Letter on Implementing the Ratio studorium and on Following the Teachings of St. Thomas, from Claudius Aquaviva, S.J., to the Provincials, 14 December 1613

A considerable time has now passed since we stated the reasons for the rules we prescribed on 24 May 1611, especially in regard to exercising care about the solidity and uniformity of our teachings. Indeed these two characteristics are so necessary and so important for the well-being of the Holy Church and for the preservation of the Society that they were recommended by Father Ignatius in the *Constitutions,* parts 3 and 4. In our letter we ordered all the Provincials in Europe to meet with some of their principal theologians, in order to consider carefully what could be done in this matter, and to send us their views, which they have done. In studying these reports we see quite clearly, and not without great personal satisfaction, that all fall into the same line and all point in the same direction. Although some of the Provincials disagreed on the specific recommendations they proposed, still either all or most agreed on many things. These included requests for more lasting rules, painful recalls to the practical life, other suggestions that are more of an outcry than a help, and finally a harsher recommendation, which is quite unusual for the Society, namely, to impose an order of obedience on our professors and censors and to constrain them with an oath that they would accept or reject various opinions.

Therefore, after having commended all this to the care of God, after many liturgies and speeches, and after mature consideration and subsequent consultation with the Assistants, we have finally decided hereby that one thing is sufficient and is to be kept in view above all the difficulties involved in this matter, namely, that our book *Ratio studiorum* is henceforth to be observed more carefully. As a result we wish to impress again in a stronger way upon the Provincials that they are to observe exactly the rules which apply to them. These rules are quite copious and clear in that book about studies, and are stated again in the *Regulae,* in the section pertaining to Provincials, chapter 6, "On Literary Studies." Next we wish to impress upon the Rectors that they carefully observe the rules that apply to them. The same applies to the prefects of studies and to the professors of the higher faculties, especially scholastic theologians. If everyone eagerly carries this out, it is to be hoped that, with the help of Christ, all will turn out quite well.

In regard to the *solidity* of our teaching, we will keep ourselves quite secure if we follow one author, namely, St. Thomas, as was prescribed by canon 9 of the Fifth General Congregation. And at the same time we will maintain *uniformity.* For it was this same double characteristic of doctrine which was considered by the General Congregation in decrees 41 and 56, on which canon 9 was based. Nor can anyone effectively say that the mind of St. Thomas is not always well known, for that decree removes all such doubts. If an opinion is contrary to St. Thomas, it is not necessary to examine it; since it is not a solid doctrine, we should never follow it. If an opinion is in agreement with the Holy Doctor, then it is equally certain that that doctrine has the two characteristics we seek. And if an opinion is doubtful because many Thomists among the old responsible teachers have interpreted it in many ways, then together with them it is permitted to accept that opinion which is thought to be more probable and not contrary to the mind of St. Thomas. But in this case be careful about the following serious warning. If one is attached to some particular doctrine, do not try to argue that it is the same as the teaching of St. Thomas by seizing upon a few words found here or there in his writings, and then contorting their meaning, so that you pass off your own opinion as that of St. Thomas. Rather the reliable doctrine and true sense is to be derived from those passages where St. Thomas expressly treats a specific topic, and not where he happens to touch upon it in a transitory way while he is dealing with something else.

Having said this, there is no need to write down a catalogue of opinions that we ought to accept or reject as probable or improbable. For if we follow one author, then all difficulties are removed. And this is not a denial of our mental powers, but only a restriction. For our talents can be used to confirm that author's doctrine with further arguments, and to illustrate and extend it more widely. And thus there indeed is a large field in which the efforts of the teacher can be exercised by following in the footsteps of the approved author. For even though there is not always a discussion of whatever topic is of present concern, let us point out that it is quite appropriate to examine the numerous and varied discussions of the sacred doctors, who have published both commentaries on the sacred writings and compendia on other important topics, and who have not deviated even a trifle from the true path that we intend.

Moreover there are those who think that the variety of opinions that are found in many writers of our Society is not a hindrance to the uniformity of doctrine, because they believe that they are free to follow any of our members whose books have been published with the support and approval of the Superiors. But we have clearly declared that, in following an opinion, our teachers and writers are not permitted to depart from St. Thomas. If an opinion has been permitted up to now in published books, as was said about matters in doubt, that opinion will be free from novelty, boldness, and error; but this does not mean that others ought to follow it if it is known to be opposed to St. Thomas.

As a result it would be quite worthwhile to call a convocation of the teachers annually, at the beginning of the school term, to read to them the letter I wrote concerning the solidity and uniformity of doctrine. It would also help if the Superiors of the colleges, where philosophy and theology are taught, would at the same time give an exhortation in which they would zealously advocate the decree concerning the doctrine of St. Thomas, and especially concerning the rules prescribed either by us or by the *Ratio studiorum,* so that nothing is overlooked by negligence.

Above all let the Provincials remember what is contained in rule 9 in the section pertaining to them in the *Ratio studiorum,* namely, that those who are not well disposed towards St. Thomas and who are prone to novelties in either theology or philosophy should not be promoted to academic chairs. Those who are of like mind and who are also actually teaching, especially if they follow opinions that are confused and not solid, should be openly removed from the office of teaching and assigned to other ministries of the Society. Whoever teaches views contrary to St. Thomas, or who introduces novelties into philosophy on his own initiative or from obscure authors, is ordered to retract such things immediately and not wait until the end of the year or of the course. For in addition to the fact that evil becomes established, and even increases, with the passage of time, such novelties also become strengthened in those teachers who have never been warned to the contrary for a whole year. Nor is ignorance any kind of a legitimate excuse for those in charge. For their office charges them to give serious consideration to this matter in many ways: (1) when they have occasion to hear teachers, according to rule 17 for their office; (2) when they repeatedly take part in petitions and in public or domestic disputations; and (3) when they interrogate students and examine occasionally the commentaries they submit. If they attend to these matters with diligence, they could not avoid recognizing those who commit errors.

As a result it will be necessary to establish censures, penalties, public and private reprimands, and serious warnings, in addition to the retraction of opinions. This is so that it will be understood that this matter is to be taken seriously, not superficially, and so that it will be pursued vigorously by the Society and its Superiors, with the result that it will command a greater importance than would otherwise be believed for the goals of the Society, for the good of the Church, and for obedience to God. Hence the Prefects must tighten the reigns rather than loosen them. If a doubt should arise, or if a teacher chooses to defend his own opinion obstinately, then the Prefect and the Rector, proceding with great authority and kindness, will consult on the matter and hand it over to other learned men of our Society for more careful examination. After this they will order the teacher to acquiesce and to let himself be governed, for we have schools, as is pleasing to God, in which anything that falls outside the vow of obedience is simply unacceptable. Likewise we warn our colleagues not to blame or attack each other. If we follow one author, then although we will disagree, which is unavoidable on a few matters, still we will be mindful that we agree honorably and respectfully on common matters.

When the Provincial makes his visits, he should investigate these issues carefully, and should apply proper and effective remedies, even to the point, as was said, of removing anyone from the classroom who will not accept the rules of the General Congregation, our orders, and indeed even those of Father Ignatius, who so strongly recommended this same solidity and uniformity of doctrine. Furthermore the Provincial should notice and make sure that the opinions that are taught in philosophy are subservient to theology, and that our philosophers follow one author, as long as that doctrine does not depart in any way from Catholic truth.

In matters not treated by St. Thomas, and therefore on which his opinion cannot be known, it is necessary to note that one must be careful to accept views that are the least removed from his principles and from the rest of his teachings. For even if St. Thomas has not treated a topic, still he would never have taught anything which is contrary to his own principles and which is not consistent with his own doctrine. And even if perchance someone were unaware of this, he still is not free to devise or present a new doctrine on his own initiative, unless it is supported by responsible and established authors. Nor do we believe that it is permissible to follow an uncommon and idiosyncratic interpretation, a practice which often and quickly causes one to rush headlong into danger and to get lost. The Prefects of Studies are under obligation to realize that for many reasons it is not helpful to have as professors people who are clearly ill-disposed towards the doctrine of St. Thomas.

If anyone in the Society adds something without permission after a book has been corrected by the censors, he is to be severely punished. This is an obligation of the gravest importance to the Society.

Finally one further point is to be noted. It is lamentable that minds are confused and the solidity of doctrine is weakened because some teachers occasionally create problems by asserting that all opinions are only probable. The result is that students do not hold anything for certain on which they can stand. Teachers should always exercise great care to assert that some opinions are solid.

Of the various topics that relate to doctrine, one is of special importance, namely, the topic of grace. On this we especially desire to establish uniformity among the members of the Society. Therefore we have formulated a decree, which you will see is added to this letter. We order that this decree be made known to all teachers and that it be zealously observed everywhere by the professors of theology.

Decree on the Uniformity of Doctrine, Especially in regard to Efficacious Grace

For the meeting of minds which is strongly recommended in the *Constitutions,* for the sake of the uniformity and solidity of our teachings, and for the good of the external reputation of the Society, it is most important, especially in grave matters, that we avoid as much as possible opportunities of devising new opinions. This has been considered by the Assistants on many occasions and for a long time, and has been most strongly recommended, with the help of God. We have decided that strong legislation and grave orders are required in this matter. As a result, by the authority and binding power of our office, and before witnesses, we have legislated and ordered the following. In dealing with the topic of divine grace, our members are to follow the opinion which has been developed by many writers of our Society in their books, lectures, and public disputations, and which has been explained and defended by our most responsible Fathers as more consistent with the views of Saint Augustine and Saint Thomas, in the controversy on the assistance of divine grace held before Pope Clement VIII, of pious memory, and before the present Pope Paul V. In the future we will in particular teach that there is a distinction between what is called sufficient grace and what is called efficacious grace, which is automatically effective. This distinction is found at the level of human actions because in the former case, but not in the latter case, the effect arises from the exercise of free will cooperating with that grace. This distinction is also found at the level of the being of things. For, given God's knowledge of conditionals, His efficacious power, and His intention of doing only what is good for us, efficacious grace purposefully chooses the means, the manner, and the time in which God foresees that the effect will infallibly occur, and it will operate otherwise if God will have foreseen that the effect will not occur. As a result, in regard to both morality and to special gifts, there is always something more contained in efficacious grace than in sufficient grace, in regard to their reality. In this way God causes us to act automatically, and not merely to be able to act because He has given His grace. The same thing is to be said about perseverence, which without doubt is a gift from God.

Christopher Scheiner's *Prodromus pro sole mobile* (1633, pub. 1651)

Book 1, Chapter 1: My Reason and Occasion for Writing

Some years ago I was the first person, which I should know, to write about sunspots, in the form of several letters which I then published for public inspection under the pseudonym of "Apelles hiding behind his painting." Someone else [i.e., Galileo] immediately cast his eyes upon the views presented and seen in these writings, and arrogantly claimed them as his own discoveries. For a long time and in various ways he criticized Apelles and his writings, and even accused him of plagiarism. And for a longer time I put up with his hateful injuries to my good reputation. Although the whole first book of my *Rosa ursina sive sol* was intended to refute him, I do not think that the entire truth of the matter has been made clear to the city, and thus to the whole world. Meanwhile Apelles's critic has not muttered a word in reply. Even though I sent him my large book, the *Rosa ursina,* there can be no doubt that he did not welcome it, nor did he examine it carefully.

If this had not been the case, he would have rightly learned something, previously unknown to him, which he can in no way challenge or disapprove, namely, that from the observations that I have accumulated in that book, I have uncovered mathematical demonstrations of two new motions of the sun. The first motion lasts twenty-seven days and some hours, and proceeds from the rising region of the solar disc to its setting. The second lasts exactly one year, proceeds from the setting to the rising, and is congruent with the celestial appearances of sunspots.

But on the other hand he would now in no way admit that he is my student, or that I am his teacher (although once in 1612 he grandly promised this in his third letter to the illustrious Mark Welser, Prefect of Augsburg). Furthermore he hoped that the credit for discovering the annual rotation of the sun would be taken from me and given to him in the mind of the ignorant common man, and perhaps also among some scholars.

In his *Dialogue* of four days, published in Florence in 1632 in Italian, he states that formerly he had indeed thought for many years that the apparent motions of all sunspots on the disc of the sun are rectilinear, equal, and parallel to the ecliptic. He also claims that later, after he had discovered a curvilinear motion of one of the sunspots observed at his villa, it suddenly occurred to him that the sun must stand still at the center of the universe, and that the earth must rotate annually around the sun on the ecliptic. As a result the observed annual motion of the sunspots is not due to the rotation of the sun on itself, but to the rotation of the earth as indicated, while the sun is taken to turn on its own axis, which is stable and inclined to the plane of the ecliptic (as in other matters he fails to really prove how that happens). And from this he claims to have derived the newest and most convincing argument for the Copernican system.

He is a man who is envious of his own glory and who is rich from the good luck of others. He has given us a patchwork of rags rather than submit his own work to public criticism. As was said above, he used my work of many years of carefully determining and solidly demonstrating the paths of the sunspots, which appears in my *Rosa ursina,* especially in book 3. From his own uncertain work on the apparent path of only one sunspot, he could in no way have established the apparent annual motions of sunspots, and as I will show later his determinations are based on fallacious arguments. And lest it be said that he has walked or stood in my footsteps, he has attributed the apparent annual motion of the sunspots to the motion of the earth (which he simply presumes and has never proven.) Furthermore he conceals the truth about the denseness of the sunspots, which the solar phenomena have necessarily brought to light, but which has derailed many in the darkness of error.

My friends have strongly urged me to mount a more fiery attack than I did before. As a result I will procede to go over again in detail the same ground that I covered before in my mathematical examinations of the motion of the earth and the stability of the sun. I will root out the sudden attack of fever in the fourth day of his *Dialogue,* so that it does not spread as an illness of the mind. I am confident that the only way for me to do this is to prove that the sun moves and that the earth is at rest, because this will serve as a medicine for the mind of those who have caught the fever. As I was proven successful in my earlier discovery of sunspots, here also I hope to please Urania [the Muse of Astronomy] when I re-echo the vows of so many very great men who were distressed at the thought that the earth truly might move. While I prepare still greater things, I send this *Prodromus* [i.e., a messenger] as a helper for true mathematicians, to protect them from deception and myself from attack, always remembering religious modesty, which seeks to know only the truth, to defend the innocent, to injure no one, and to bear any wrong.

Therefore in this work, the *Prodromus,* I will not attack all the devices and techniques that the author of the *Dialogue* uses to show the stability of the sun and the motion of the earth. Rather I will attack only his new and his principal fortress, as he admits, and I will capture and tear it down to the ground, armed with the truth, which is always victorious. Indeed the central basis of his case is that the apparent annual motion of sunspots is not to be judged as relative to the globe of the sun but to the mass of the earth. I will use a double attack against this fortress of his; the one positive, the other negative. With the former I will defend myself and smash the force of his assault on the truth; with the latter I will level to the ground all his defenses and fortifications.

As the reader will see, I will sharply distinguish between the tri-monthly spots which have a triple period around the sun, the periodic monthly paths of the sun itself, and the sun's annual revolution. And I do not presume but directly prove this (as will be clear in book 2 of the *Prodromus*) with a method that is used by true astronomers and that is needed for effective persuasion.

The dialogue method does not do this, but merely begs the question unless it is nourished at another table. He does not prove that the earth moves and that the sun is at rest from the past and present appearances of sunspots (which he had promised he would do). Rather from his presupposition of the motion of the earth and the rest of the sun (which was to be proven), this divinely inspired man (which will be clearly shown later from his own words) predicts the future appearances of the sunspots by using my data on their past appearances. As a result he infers that the earth moves from the motion of the earth, and that the sun is at rest from its being at rest, thus mixing everything together with obvious fallacies, empty guesses, and pure divinations, like the wanderings of the Egyptians. Now that he has publicly presented such things to the world, what value can this method be judged to have?

Every sunspot that he reports, except for only one (which will be examined more closely in its proper place), was observed not by him but previously by others. This then, I say, is merely a bare report, for he gives us only his words to read, which say that something has been observed, but beyond the words he offers nothing worthy of belief. He does not even explain the notion of observation or vision. He does not discuss the moments of time, or the locations of apparent places on the disc of the sun which are excluded, nor does he investigate the true places of motion and rest. He is silent about an infinite number of other things, which arise later from his writings, and which are quite necessary to establish his credibility. Indeed this supercilious academician has proposed that the reader simply pay attention to what he has written as sufficient in itself, without using his own eyes or his own brain.

That messenger was once thought to be above the stars, but he has fallen down to us from the moon. At one time he noticed that a single sunspot, observed among some others, had a curved path. From this he then proclaimed that the sun is at rest and the earth moves. Hence the earth actually does move, and the sun really is at rest, because that lunatic thinks so, and has proclaimed that in his book. But that messenger has not proven these asssertions. Nothing is established about the way things are just because that herald of the heavens wishes that it be so, substituting his will for reason. New things revealed by the heavens in this era have taken hold of human fancy. But it is not scientific to philosophize by willing rather than by thinking. This Mercury gives us nothing beyond his words. Let the reader listen carefully and with prudence.

CHAPTER ONE. THE LEGAL CASE AT GALILEO'S TRIAL

1. For a discussion of these topics, see Blackwell (1991).

2. For a full translation of Foscarini's *Letter* and related documents, see Blackwell (1991, appendices 6–9).

3. The members of this Special Commission were the Master of the Sacred Palace, Niccolò Riccardi, O.P., the papal theologian Msgr. Agostino Oreggi, and Fr. Melchior Inchofer, S.J. When the Commission was reconvened in April of 1633 for a second opinion during the trial, Riccardi was replaced by the Theatine priest Fr. Zaccaria Pasqualigo. All were trained and experienced in theology, but had little or no background in science or astronomy.

4. For a translation of the key documents in the Galileo trial, see Finocchiaro (1989, 256–93).

5. For an example, see Farinacci (1616).

6. The Latin phrase used here is *ex suppositione,* which at that time meant that Copernicanism is not the true account of the world, but it would be harmless if one were to assume it counterfactually for purposes of mathematical computations in astronomy. It did not mean "hypothetically" in the contemporary scientific sense of an assumption which is possibly true about the world and for which more evidence is to be sought for verification. Hence Bellarmine also asserted that Copernicanism is not true of the real world absolutely.

7. For translation of this letter, see Blackwell (1991, appendix 8).

8. For these reports, see Finocchiaro (1989, 262–76).

9. Maculano's letter to Cardinal Barberini, 22 April 1633. This letter was discovered after the 1998 opening of the archives of the Congregation of the Doctrine of the Faith (the successor to the Holy Office and the depository of its files), and was first published in 2001. For the Italian text of the letter, and a detailed explanation of its significance, see Beretta (2001a).

10. We know from a letter to Galileo from Castelli in October 1632 that Maculano agreed with Galileo on Copernicanism and also thought that such questions should not be settled by appeal to scripture. Of course, the trial was really not about the Bible, however, but about authority and obedience to church decrees.

11. Maculano's letter to Cardinal Barberini, 28 April 1633, in Finocchiaro (1989, 276–77). It should be noted that this letter was not discovered until 1875. Without it, the dramatic change in the atmosphere of the trial from the first to the second session is simply inexplicable. There would otherwise be no rational reason why Galileo would have made a confession in the second session after the good case he had made for himself in the first session. But the intervening plea bargain makes this shift quite understandable.

Annibale Fantoli (2003, 314–16) has recently given an interpretation of Maculano's letter that is quite similar to mine. He suggests reasonably that the authority to deal with Galileo extrajudiciously after the first interrogation was likely given to Maculano by Pope Urban VIII through his nephew Cardinal Francesco Barberini, who at the time was a member of the Congregation of the Holy Office.

12. For the full report, see Finocchiaro (1989, 281–86).

13. For a fuller account, see Blackwell (1991, 64–76).

14. For a translation of Galileo's Letter to Castelli (including notations of Lorini's changes in the text), see Blackwell (1991, appendix 4). Following the manuscript researches of Mauro Pesce (1992), Annibale Fantoli (1996, 133–34) has argued that the true original text of the Letter to Castelli is the Lorini version, which Galileo later amended to soften its language in the version which he sent to Msgr. Dini as his actual set of views. There is no need for us to adjudicate here between these two interpretations. For in either case the important point for us is that the author(s) of the summary report began by recounting at some length the complaint brought by Lorini to the Holy Office many years earlier, which, as they must have known from their own file on Galileo, had already been dismissed in 1615 as unsubstantiated.

15. Giorgio de Santillana (1955, 308) has argued that the two chief authors of the report were the Proctor Fiscal, Fr. Carlo Sinceri (the courtroom interrogator at the trial) and Msgr. Paolo Fabei, the Assessor (an office intermediary between the Commissary General and the cardinal members of the Congregation of the Holy Office). Handwriting analysis shows that the summary report was written by the same unknown clerk who wrote the depositions, but that person would have merely been taking dictation.

16. The other two cardinals who did not sign were Gasparo Borgia and Laudivio Zacchia. Many Galileo scholars have interpreted the absence of Cardinal Francesco Barberini's signature as a sign of his personal opposition to the conclusion of the trial. But quite recently it has been pointed out by Fantoli (2003, 541) and Beretta (2001c, 568, 572) that that may not be the case since (1) such documents at that time often did not include a full set of signatures and (2) both Borgia and Barberini were preoccupied that day at a papal audience concerning Spanish objections to papal policy on the conduct of the Thirty Years War, then in progress.

17. Since no human can know the mind of another person, the standard procedure of the Holy Office was not to pronounce that a defendant was a heretic, but only that he was held in "suspicion of heresy." There were three technical categories of suspicion: light, vehement, and violent. Galileo's offense was thus judged to be in the middle category of severity. It could have been worse.

18. The prohibition was removed in 1822, and the *Dialogue* did not appear in the next edition of the *Index* (1835).

Chapter Two. Melchior Inchofer's Role in the Galileo Affair

1. For a strong indictment of the Jesuits for plotting against Galileo, see Santillana (1962, chapter 14). On the other hand, for a vigorous denial that the Jesuits were enemies of Galileo, see Gallo (1963, appendix 4, 93–114).

2. According to Pietro Redondi (1987), Orazio Grassi was the main person responsible for Galileo's trial. Redondi's thesis, which has not been widely accepted, is that Grassi was the author of an anonymous complaint to the Holy Office against Galileo's atomism as conflicting with the Catholic teaching on the Eucharist, and that the trial over his Copernican views was only a diversion from these other more serious charges.

3. On this policy, see Blackwell (1991, chapter 6).

4. For a balanced and more detailed assessment of this issue, see Fantoli (1994, 428–31).

5. For example, Inchofer's name is not listed either in the biographical appendix or in the index of Drake (1978).

6. The following account is based partially on biographical information on Inchofer generously supplied in 1989 by Fr. Francis Edwards, S.J., then the archivist at the Archivum Romanum Societatis Iesu in Rome.

7. The full title of the original treatise was *Epistolae B. Virginis Mariae ad Messanenses Veritatis vindicata ac plurimis gravissimorum scriptorum testimoniis et rationibus erudite illustrata* (1629). The title of the second, revised edition was *De epistola B. Virginis Mariae ad Messanenses conjectatio plurimis rationibus et verisimilitudinibus locuples*.

8. In one of them, his 1635 *Historia sacrae latinitatis, hoc est: de variis linguae latinae mysteriis*, he argued for the odd claim that Latin was the language spoken by Christ and by the blessed in heaven!

9. Did he do this to soften the political pressures on himself; to aid Galileo personally, who was an old friend and a fellow Tuscan; to defer to the Duke of Tuscany, who was on Galileo's side; to protect Galileo from the more serious charges of Eucharistic heresy, as Redondi had claimed? In the absence of any specific documentation, we simply do not know why Urban VIII chose this course of action.

10. The exception is Pietro Redondi (1987, 244–49), who argues that the purpose of the Special Commission was to save Galileo from a trial over his supposedly heretical views about the Eucharist by finding some other, much less serious set of charges.

11. This sentence reads: "In conformity with the order of Your Holiness, we have drawn up an account of the entire series of events which occurred regarding the publication of Galileo's book, which later was published in Florence" (Galileo 1890–1909, 19:324).

12. Redondi (1987, 249–55) argues that the third member was a Theatine priest, Zaccaria Pasqualigo, who indeed was a member of the Special Commission when it was called upon a second time in April of 1633 for another report on the *Dialogue*. We will examine Pasqualigo's contributions later.

13. This suggestion actually had originated with Campanella himself, who described the Special Commission as a group of "irate theologians" (Galileo 1890–1909, 14:373).

14. Campanella's book to which Riccardi refers is his *Apologia pro Galileo*, which was placed on the *Index* immediately after its publication in 1622. For an English translation, see Campanella (1994).

15. See Galileo (1890–1909, 19:324–30). A full English version can be found in Finocchiaro (1989, 218–22).

16. For a detailed discussion of these events relating to the permissions, see Fantoli (1994, 312–21).

17. For Bellarmine's letter, see Galileo (1890–1909, 12:171–72). For English translations, see Blackwell (1991, 265–67) and Finocchiaro (1989, 67–69).

18. William R. Shea (1984, 287) has argued that the Special Commission itself found this memo because it had direct access to the files of the Holy Office. Fantoli (1994, 441–42) thinks that that is not likely because of the strict secrecy rules at the Holy Office, and that the memo must have been given to the Commission by some unidentified member of the Holy Office who was hostile to Galileo. Either way this decisive memo made its first appearance in the Galileo case as a result of the investigations of the Special Commission.

19. See Galileo (1890–1909, 19:321–22). Galileo, of course, never saw this memo, which was kept in the secret files of the Holy Office. As discussed in the previous chapter he relied on another account of the meeting contained in a letter he later requested from Bellarmine. That letter (Galileo 1890–1909, 19:348; Blackwell 1991, 127) said that the order merely was that Copernicanism "cannot be defended or held." The words "teach" and "in any way whatever, orally or in writing" were not included in Bellarmine's account. That would have made the publication of the *Dialogue* much less objectionable.

20. Stillman Drake (1979, 339–40) has argued that this change of heart occurred when the pope learned of the injunction of 1616 just discovered by the Special Commission.

21. Riccardi was replaced by a young Theatine priest, Fr. Zaccaria Pasqualigo, who was a theologian, not a scientist.

22. Oreggi's report was dated 17 April. The other two reports are not dated, but must have been submitted on or about the same date.

23. Inchofer recognized this point at the beginning of his report, and used it to strengthen his argument. "Therefore all those reasons on which Galileo—in a categorical, absolute, and nonhypothetical manner—grounds the earth's motion necessarily also prove or assume the immobility and central position of the sun" (Finocchiaro 1989, 263).

24. See Blackwell (1991, 112–22). This same distinction is also found in Tommaso Caccini's complaint to the Holy Office against Galileo on 20 March 1615 (Galileo 1890–1909, 19:307–11).

25. His reference is to Augustine's commentary on Psalm 118, section 17.

26. See Finocchiaro (1989, 265). This refers to Scheiner's *Rosa ursina sive sol* (1626–30), which begins with a very lengthy and bitter attack against Galileo.

27. For a detailed examination of this point, see Baldini (1992, 19–73).

28. For example, he went out of his way to criticize Galileo's views, contained in a work written almost twenty years earlier (the *Letter to the Grand Duchess Christina*), that the Bible speaks to the level of the common person, adding, "Then he ridiculed those who are strongly committed to the common scriptural interpretation of the sun's motion as if they were small-minded, unable to penetrate the depth of the issue, half-witted, and almost idiotic" (Finocchiaro 1989, 263).

29. The full title of Inchofer's book is *Tractatus syllepticus, in quo, quid de terrae solisque motu vel statione, secundum S.Scripturam et Sanctos Patres sentiendum, quave certitudine alterutra sententia tenenda sit, breviter ostenditur* (1633). The book must have been completed no later than mid-August of 1633 because the two evaluations of the book recommending publication are dated 18 and 22 August. The imprimatur was granted by Fr. N. Riccardi, O.P., Master of the Sacred Palace, who had figured so prominently in the dispute over the permissions to publish Galileo's *Dialogue*.

CHAPTER THREE. THE SCRIPTURAL CASE AGAINST COPERNICANISM IN 1633

1. The unpublished manuscript is in Rome in the Biblioteca casanatense, MS 182.

2. For the documents from which these two points are taken, see Beretta (2001b, 321–22).

3. Francesco Beretta (2001b, 324) has concluded, "Ce fut donc à Melchoir Inchofer qu'on confia la tâche de fournir une justification doctrinale de la condemnation de Galilée."

4. See Blackwell (1991, appendix 6) for a full English translation of this booklet.

5. For a discussion of this pivotal controversy, see Blackwell (1991, chapter 4).

6. In his acceptance of the report of the Galileo Commission on 31 October 1992, Pope John Paul II (1992) admitted that theological mistakes were made in the Galileo case, but he did not identify the theologians who made the mistakes. Bellarmine and Inchofer are strong candidates for inclusion on this list for the reasons indicated in this paragraph.

7. See especially chapter 11.

8. The Latin term used here and later is *probabilius,* which is the comparative form of the adjective, i.e., "more probable" or "rather probable."

9. See especially the last paragraph of *Tractatus,* chapter 10 (pp. 152–53 above), which is the culmination of Inchofer's main argument.

10. The Jesuit philosophers at the time were certainly aware of the basic logical relations of opposition that can occur between two propositions, since that had long ago been worked out by Aristotle in his *De interpretatione* and was widely taught in the Jesuit schools. The relevant types of *logical opposition* here are (1) *contraries* (e.g., "All swans are white" and "No swans are white") in which either one could be true, but not both; (2) *sub-contraries* (e.g., "Some swan is white" and "Some swan is not white") in which either one could be false, but not both; and (3) *contradictories* (e.g., "All swans are white" and "Some swan is not white") in which one must be true and the other must be false. The question would be which type of opposition holds between heliocentrism and Aristotelian geocentrism, or between heliocentrism and biblical geocentrism (which was Inchofer's own point of view). The logical opposition between more than two views (which was the situation at hand) becomes more complex, and especially if one or more view is stated in compound sentences, as the Holy Office tried to give for the Copernican theory. To say the least, their decision was logically ambiguous.

CHAPTER FOUR. CHRISTOPHER SCHEINER'S DILEMMA

1. This is now the city of Nysa in far southwestern Poland about 125 miles east-northeast of Prague.

2. For analyses of the sunspot dispute, see Dame (1968); Shea (1970); Sharratt (1994, 98–106); and Fantoli (1994, 132–41, 156–60).

3. For an excellent account of Clavius's scientific work, see Baldini (1983).

4. For a clear analysis of this argument, see Shea (1970, 504–6).

5. It is clear from the title page that Scheiner was the author of this book, although that credit is given to his student Joannes Georgius Locher, in a weak subterfuge. Frequent and praising references are made to Galileo's telescopic discoveries. In a maneuver soon to become common in Jesuit books on astronomy, all of the major contending theories are carefully explained and diagrammed for the reader's information. Copernicus's sun-centered model is rejected for the usual physical reasons, primarily that it requires a truly immense increase in the size of the universe. The author claims that it is not yet possible to decide conclusively between the various earth-centered models, but much attention and appeal is made to Tycho Brahe's version of geocentrism.

6. For further discussion of these issues, see Blackwell (1991, 48–51).

7. For an explanation of this point, see Blackwell (1991, 143–47).

8. It is interesting to note in passing that, although Galileo clearly states his commitment to Copernicanism in two places in his *Letters on Sunspots* of 1613, *neither* pas-

sage was challenged by the ecclesiastical censors. However there were two versions of a passage near the end of his Second Letter to Welser that stated in effect that the notion of the alterability of the heavens conforms well with the scriptures. *Both* of these texts were challenged, and were finally replaced with a comment to the effect that Aristotle himself would have changed his mind about the inalterability of the heavens if he had seen the new evidence about sunspots. In 1613 it was this Aristotelian doctrine, and not yet Copernicanism, that apparently was of primary concern in regard to religious orthodoxy. For the wording of all three versions of the disputed passages, see Galileo (1890–1909, 5:138–40).

9. The second pseudonym was added by Scheiner at the end of his *Accuratior disquisitio* (published in 1612), which was a more detailed follow-up of his original *Letters on Sunspots* earlier that year.

10. The old story (found in Pliny the Elder, *Historia naturalis,* bk. 35, chap. 36) is that a cobbler criticized the shoes in one of Apelles's paintings, and returned later to find them repainted. Emboldened, the cobbler proceeded to criticize the shape of the thigh, whereupon Apelles appeared from behind the painting and said, *Ne sutor ultra crepidam* ("Let not the cobbler go beyond the shoe"). Erasmus rendered this later into the proverb, "Cobbler, stick to your last."

In a later letter to Welser (25 July 1612) (Galileo 1890–1909, 5:57) Scheiner explicitly identified his own drawings of sunspots with "Apelles's painting": "Various people have had different understandings of the sunspots which I have drawn on Apelles's painting."

11. This refers to Ajax the Greater, son of Telamon, who used a famous shield of seven layers of oxhides covered by a sheet of bronze. After the death of Achilles at Troy, the latter's armor was awarded to Ulysses rather than to Ajax, who thereby was so upset that he became mad and took his own life.

12. For these studies see Baldini (1992), Dear (1987), and Feldhay (1995).

13. The best known instance of this widely used advice is found in Bellarmine's Letter to Foscarini, 12 April 1615 (Galileo 1890–1909, 12:171–72).

14. Rivka Feldhay takes this to be the main cause, tracing its influence back to decisions made in 1599 that defined the character of the Jesuit school curriculum and educational policy. "Nonetheless, from the point of view of the history of Western science the year 1599—the date of publication of the last *Ratio studiorum*—was the lost moment of the Jesuit educational program" (Feldhay 1995, 232). If this be true, then the die was already cast before Scheiner was on the scene, and certainly before the condemnation of Copernicanism in 1616.

15. For detailed analyses of the controversy over the comets of 1618, see Fantoli (1994, chapter 4) and Drake (1978, 267–80, 283–88). The treatises published on both sides of the debate are in Galileo (1890–1909, vol. 6) and in English translation by Drake and O'Malley (1960).

16. Giovanni Remo's Letter to Galileo, 24 August 1619 (Galileo 1890–1909, 12:489).

17. This is from the third paragraph of the *Assayer.* The translation is by Drake (1957, 232).

18. Drake (1957, 232n) suggests that Galileo's target here may have been Jean Tarde.

19. It is indeed ironic that this celebration took place just a few days before Aquaviva's letter of 24 May 1611 (see appendix 2.B) prescribing "solid and uniform teaching" for the Jesuits.

20. See Blackwell (1991, 150–52).

21. For a discussion of this Jesuit notion of religious obedience, see Blackwell (1991, chapter 6).

22. This is from Grassi's *Libra* (Galileo 1890–1909, 6:151). The translation is from Drake (1978, 278). The Latin text is (intentionally?) ambiguous, and is translated with the opposite meaning by M. Sharratt (1994, 135) as, "I wish everyone to witness that here the least of my wishes is to contend for Aristotle's opinions."

23. My efforts to find in the Jesuit archives in Rome the specific reason for Scheiner's call to Rome have not been successful.

24. Galileo's Letter to Marsili, 19 April 1629 (Galileo 1890–1909, 14:36; translation is from Fantoli 1994, 345–46). In his second edition, Fantoli (1996) gives a more detailed account of how relations between Galileo and Scheiner deteriorated between 1613 and 1632.

25. Both Galileo and Scheiner thought of the sun as a solid body; hence for them its period of rotation on its axis would be the same at all latitudes. Actually, however, since the sun is not a solid but rather a gaseous plasma, its period of rotation varies from about twenty-five days at the equator to about thirty-five days in its polar regions.

26. Scheiner's measurement was 7 degrees, 30 minutes. The modern value is 7 degrees, 15 minutes.

27. The apparent curvature of the path of sunspots over time is due to two facts: (1) the inclination of the axis of the sun to the perpendicular to the ecliptic is 7.25 degrees, and (2) the inclination of the axis of the earth to the perpendicular to the ecliptic is 23.5 degrees. As a result the relative motion of these two axes over one year causes a change in perspective for an observer on the earth, and hence the paths of the sunspots appear to that observer to change curvature over the course of the year.

28. For examples, see Stelluti's Letter to Galileo, 10 January 1626 (Galileo 1890–1909, 13:300); Descartes' Letter to Mersenne, 4 March 1634 (Galileo 1890–1909, 16:56); and Peiresc's Letter to Gassendi, 10 September 1633 (Galileo 1890–1909, 15:255).

29. For these censors' reports, see Baldini (1992, 100–101). Baldini sees these reports as attempts to free Jesuit science from the imposed restriction to Aristotle. See also McColley (1941, 63–69).

30. Castelli's Letter to Galileo, 9 June 1632 (Galileo 1890–1909, 14:360; the translation is from Fantoli 1994, 371).

31. Near the end of his *Dialogue* Galileo lists the sunspot argument as second among the three strong grounds for Copernicanism, mentioning it even before his own unique argument from the tides.

32. For a justification of this point in Scheiner's own words, as well as a representative sample of his hostile attitude towards Galileo, see the first chapter of his *Prodromus,* which is translated in appendix 3.

33. These passages can be found in Stillman Drake's translation of Galileo's *Dialogue* (Galileo 1967, 52–55, 345–55).

34. For the debate over this point, see Drake (1978, 309–11, 332–35) and Fantoli (1994, 308–10, 345–48).

35. Scheiner's Letter to Kircher, 16 July 1633 (Galileo 1890–1909, 15:184).

36. Peiresc's Letter to Gassendi, 10 September 1633 (Galileo 1890–1909, 15:254).

37. As Ugo Baldini has suggested to me, there may be no such evidence because the decision may have been communicated to Scheiner orally rather than in writing, since he lived in Rome at the time. We have been able to determine, however, that during his last years at Nissa, he asked the Jesuit General in Rome on several occasions for permission to publish the *Prodromus,* but was rejected each time without explanatory comments in the correspondence.

BIBLIOGRAPHY

Baldini, Ugo. 1983. "Christoph Clavius and the Scientific Scene in Rome." In *Gregorian Reform of the Calendar: Proceedings of the Vatican Conference to Commemorate Its 400th Anniversary, 1582–1982*, ed. G.V. Coyne, S.J., M.A. Hoskin, and O. Pedersen, 137–69. Vatican City: Vatican Observatory Publications.

———. 1992. *Legem impone subactis: Studi su filosofia e scienza dei gesuiti in Italia, 1540–1632*. Rome: Bulzoni editore.

Beretta, Francesco. 1999. "Le procès de Galilée et les archives du Saint-Office. Aspects judicìaires et théologique d'une condemnation célèbre." *Revue des sciences philosophiques et théologiques* 83:441–90.

———. 2001a. "Un nuove documento sul processo di Galileo Galilei. La lettere di Vincenzo Maculano del 22 Aprile 1633 al Cardinale Francesco Barberini." *Nuncius* 16:629–41.

———. 2001b. "Omnibus Christianae Catholiocaeque Philosophiae amantibus, D.D. Le *Tractatus syllepticus* de Melchior Inchofer, censeur de Galilée." *Freiburger Zeitschrift für Philosophie und Theologie* 48:301–25.

———. 2001c. "Urban VIII Barberini protagoniste de la condemnatione de Galilée." In *Largo Campo di Filosofare: Eurosymposium Galileo 2001*, ed. Jose Montesinos and Carlos Solis, 549–73. La Orotava, Spain: Fundación Canaria Orotava de Historia de la Ciencia.

Blackwell, Richard J. 1991. *Galileo, Bellarmine, and the Bible*. Notre Dame, IN: University of Notre Dame Press.

———. 1998. *Science, Religion, and Authority: Lessons from the Galileo Affair*. Milwaukee, WI: Marquette University Press.

Campanella, Thomas. 1994. *A Defense of Galileo*. Trans. R.J. Blackwell. Notre Dame, IN: University of Notre Dame Press.

Cerbu, Thomas. 2001. "Melchior Inchofer, 'Un homme fin & ruse.'" In *Largo Campo di Filosofare: Eurosymposium Galileo 2001*, ed. Jose Montesino and Carlos Solis, 587–611. La Orotava, Spain: Fundación Canaria Orotava de Historia de la Ciencia.

Cornford, Francis M. 1937. *Plato's Cosmology: The* Timaeus *of Plato.* London: Routledge & Kegan Paul.

Dame, Bernard. 1968. "Galilée et les taches solaires (1610–1613)." In *Galilée: Aspects de sa vie et de son oeuvre,* ed. S. Delorme, 186–251. Paris: Presses Universitaires de France.

Dear, Peter. 1987. "Jesuit Mathematical Science and the Reconstruction of Experience in the Early Seventeenth Century." *Studies in History and Philosophy of Science* 18:133–75.

———. 1995. *Discipline and Experience.* Chicago: University of Chicago Press.

Drake, Stillman. 1957. *Discoveries and Opinions of Galileo.* New York: Doubleday.

———. 1978. *Galileo at Work.* Chicago: University of Chicago Press.

Drake, Stillman, and C. D. O'Malley. 1960. *The Controversy on the Comets of 1618.* Philadelphia: University of Pennsylvania Press.

Fantoli, Annibale. 1994. *Galileo: For Copernicanism and for the Church.* Translated by George V. Coyne, S. J. Vatican City: Vatican Observatory Publications. (Original Italian edition, 1993)

———. 1996. 2nd ed. of Fantoli (1994).

———. 2003. 3rd ed. of Fantoli (1994).

Farinacci, Prospero. 1616. *Tractatus de haeresi.* Rome: Andreae Phei.

Feingold, Mordechai. 2003. *Jesuit Science and the Republic of Letters.* Cambridge, MA: MIT Press.

Feldhay, Rivka. 1995. *Galileo and the Church: Political Inquisition or Critical Dialogue?* Cambridge: Cambridge University Press.

Finocchiaro, Maurice A. 1989. *The Galileo Affair: A Documentary History.* Berkeley: University of California Press.

Galileo [Galilei, Galileo]. 1890–1909. *Le opere di Galileo Galilei.* Edizione Nazionale, cura et labore A. Favaro. Florence: G. Barbèra.

———. 1967. *Dialogue Concerning the Two Chief World Systems.* Trans. Stillman Drake. Berkeley: University of California Press.

Gallo, Salvatore, S. J. 1963. "Galileo fu vittima dei gesuiti?" In *Il processo di Galileo,* ed. Filippo Soccorsi, appendix 4, 93–114. Rome: Edizioni La Civiltà Cattolica.

Gorman, Michael John. 1996. "A Matter of Faith? Christopher Scheiner, Jesuit Censorship, and the Trial of Galileo." *Perspectives on Science* 4:283–319.

Grant, E. 1994. *Planets, Stars and Orbs: The Medieval Cosmos, 1200–1687.* Cambridge: Cambridge University Press.

Inchofer, Melchior. 1633. *Tractatus syllepticus, in quo, quid de terrae solisque motu vel statione, secundum S. Scripturam et Sanctos Patres sentiendum, quave certitudine alterutra sententia tenenda sit, breviter ostenditur.* Rome: Ludovicus Griganus.

———. 1645. *Monarchia solipsorum.* Venice.

John Paul II, Pope. 1992. "Lessons of the Galileo Case." *Origins* 22 (no. 22, 12 November): 371–74.

Langford, James. 1966. *Galileo, Science, and the Church.* New York: Desclee.

Lattis, James M. 1994. *Between Copernicus and Galileo*. Chicago: University of Chicago Press.

Locher, Joannes Georgius. 1614. *Disquisitiones mathematicae de controversiis et novitatibus astronomicis*. Ingolstadt: Ederianus et Elisabetha Angermaria.

McColley, Grant. 1941. "Ch. Scheiner and the Decline of Neo-Aristotelianism." *Isis* 32:63–69.

McMullin, Ernan. 1998. "Galileo on Science and Scripture." In *The Cambridge Companion to Galileo,* ed. Peter Machamer, 271–347. Cambridge: Cambridge University Press.

———, ed. 2005. *The Church and Galileo*. Notre Dame, IN: University of Notre Dame Press.

Nevue, Bruno, and Pierre-Noel Mayaud. 2002. "L'affaire Galilée et la Tentation Inflationniste à propos des notions d'hérésie et de magistère impliquées dans l'affaire." *Gregorianum* 83:287–311.

Pesce, Mauro. 1992. "Le redazioni originali della Lettera Copernicana di G. Galilei a B. Castelli." *Filogia e Critica* 17:394–417.

Quine, W.V.O. 1961. "Two Dogmas of Empiricism." In *From a Logical Point of View,* 20–46. New York: Harper & Row.

Randles, W.G.L. 1999. *The Unmaking of the Medieval Christian Cosmos, 1500–1760*. Aldershot: Ashgate.

Redondi, Pietro 1987. *Galileo Heretic*. Trans. R. Rosenthal. Princeton, NJ: Princeton University Press.

Santillana, Giorgio de. 1962. *The Crime of Galileo*. New York: Time Incorporated.

Scheiner, Christopher, S.J. 1626–30. *Rosa ursina sive sol*. Bracciani: A Phaeum.

———. 1651. *Prodromos pro sole mobile et terra stabili, contra academicum florentinum Galilaeum a Galilaeis*. [Published posthumously by the Jesuits without naming the city or the publisher.]

Sharratt, Michael. 1994. *Galileo: Decisive Innovator*. Oxford: Blackwell.

Shea, William R. 1970. "Galileo, Scheiner, and the Interpretation of Sunspots." *Isis* 61:498–519.

———. 1984. "Melchior Inchofer's *Tractatus syllepticus*: A Consultor of the Holy Office Answers Galileo." In *Novità celesti e crisi del sapere,* ed. P. Galluzzi, 283–92. Florence: Giunti Barbèra.

Shea, William, and Mariano Artigas. 2003. *Galileo in Rome: The Rise and Fall of a Troublesome Genius*. Oxford: Oxford University Press.

INDEX TO THE TRANSLATIONS
(A P P E N D I X E S   1 — 3)

R I C H A R D   J .   B L A C K W E L L

is professor emeritus of philosophy, Saint Louis University.
He is the author of numerous books, including *Galileo, Bellarmine,
and the Bible* (Notre Dame Press, 1991), and is the translator of *A Defense
of Galileo, the Mathematician from Florence* (Notre Dame Press, 1994).